INTESTINAL ABSORPTION OF METAL IONS AND CHELATES

INTESTINAL ABSORPTION OF METAL IONS AND CHELATES

By

H. DeWAYNE ASHMEAD, Ph.D.

DARRELL J. GRAFF, Ph.D.

HARVEY H. ASHMEAD, Ph.D.

With an Introduction by

Gene W. Miller, Ph.D.

Utah State University at Logan

CHARLES C THOMAS • PUBLISHER

Springfield • Illinois • U.S.A.

Published and Distributed Throughout the World by

CHARLES C THOMAS • PUBLISHER
2600 South First Street
Springfield, Illinois 62717

ISBN 0-398-05074-3

Library of Congress Catalog Card Number: 84-8621

Printed in the United States of America
Q-R-3

Ashmead, H. DeWayne.
Intestinal absorption of metal ions and chelates.

Bibliography: p.
Includes index.
1. Intestinal absorption. 2. Metal ions--Metabolism.
3. Chelates--Metabolism. I. Graff, Darrell J.
II. Ashmead, Harvey H. III. Title. [DNLM:
1. Intestinal Absorption. 2. Minerals--metabolism.
3. Chelating Agents--metabolism. 4. Biological Transport. WI 402 A827i]
QP165.A78 1985 612'.392 84-8621
ISBN 0-398-05047-3

PREFACE

IT is not often that the origin of a book can be traced to a single question, but that is the case here. We had been working with minerals that were chelated with amino acids for several years and had noted the physiological responses of thousands of animals and birds when they were administered this particular type of chelated mineral. We had also conducted countless experiments *in vitro* demonstrating the metabolic response of various tissues and organs to these minerals. That research was, and still is, important because a majority of the essential minerals consumed by man and animals is in the form of chelates, not inorganic metal salts.

One day the question was asked, "How are these chelated minerals absorbed in the intestines?" No one knew the answer.

Our studies had taught us that these chelates we had been working with were absorbed by an active transport process, but was it the same transport mechanism used for the intestinal uptake of cations? Why was the absorption of divalent mineral ions enhanced when they were each chelated with two amino acids prior to their presentation to the gut? Were the chelates absorbed intact into mucosal cells or were they dissociated in the gut before uptake could be accomplished?

The answers to these and other equally perplexing questions necessitated several new studies. In some cases we had to develop new techniques in order to obtain needed answers. In other instances we were fortunate in that we were able to modify existing procedures and adapt them to our requirements.

We have received advice and assistance from many people as we explored new biochemical and physiological ground. To these people, who are too numerous to name, we again express our appreciation. Nevertheless, we are particularly grateful to Professor Gene W. Miller, Dean of the Biology Department of Utah State University, for not only his critique of our manuscript, but also his introduction to this book. To Professor Sen-Maw Fang of the School of Pharmacy at the University of Utah who reviewed this manuscript and also contributed some of the research data in it, we express a debt of gratitude. To the former Chairman of the School of Pharmacy at the University of Utah, Professor Robert V. Peterson, and to Dr. Hsin-Hung Hsu and Dr. Dee Oyler, both of Albion Laboratories, Inc., as is Evan Jones, all of whom received and commented on this manuscript, we say thank you. And finally, a special thanks to Mrs. Laurel Moore for the typing and retyping of this manuscript.

We would like to acknowledge the most generous financial support given by Albion Laboratories, Inc. of Clearfield, Utah who underwrote the entire expense of this project.

INTRODUCTION

THE implications of the advantages of absorption of dipeptide-like amino acid chelates are far reaching in achieving a more rapid and higher blood level of minerals in animals. Studies suggest such absorption allows efficient translocation and utilization of the minerals throughout the body allowing the genetic potential to be reached with the presence of maximum functional levels of minerals.

The greater part of copper in herbages is found in the complexed state. Animals fed this complexed copper compared to copper sulfate had a much more rapid response and a greater copper storage. Similar increases are found with zinc when ingested with a casein extract. Ingestion of iron and absorption into the lumen is shown to be enhanced when histidine is present.

These phenomena and others may be explained by peptides being absorbed across the mucosal membrane at a special site. Minerals chelated to amino acids to form a dipeptide-like molecule would follow the same absorption site as dipeptides providing an accelerated route for mineral uptake separate from ionic absorption sites. That an intact chelate having peptide characteristics may also be translocated along a peptide absorption pathway is evident from experimental data presented. Ashmead, *et al.* conclude that ionic uptake into the mucosal cells requires a pump (similar to the sodium-potassium pump) and is a two-step process for active uptake. However, the dipeptide-like amino acid chelate depends on a single series of enzymatic reactions for absorption. The chelate also avoids brush border hydrolysis in the intestines and is largely transported across the mucosal cell to the plasma still intact as the original chelated molecule.

Intestinal Absorption of Metal Ions and Chelates introduces the reader to mineral needs of the body and outlines mechanisms of transport and absorption. Detailed descriptions of processes involving uptake into the intestines are presented and experimental evidence for the uptake of dipeptide-like amino acid chelates of minerals are documented. Ashmead, *et al.* discuss the advantages of absorption of both chelates and free mineral ions by the intestines. This presentation of a second absorption pathway for minerals chelated by naturally occurring amino acids is convincing and a major contribution. Such chapters make the book valuable as a reference for scientists in research and teaching as well as an introductory text for students.

Intestinal Absorption of Metal Ions and Chelates is thought-provoking and is a treatise on mineral absorption into the body. It compares mineral ions and minerals chelated by natural ligands and their absorption into the intestines.

This is a truly readable book and presents the thoughts of a group of scientists who have been able to study, assimilate and present many secondary sources as well as the most important primary sources.

Gene W. Miller, Ph.D.
Utah State University
Logan, Utah

CONTENTS

Page

Preface v
Introduction — ***Gene W. Miller*** vii

Chapter

1. Mineral Needs of the Body 3
2. Factors Affecting Mineral Absorption from the Intestine 13
3. Structure of the Small Intestine 30
4. Fundamental Aspects of Chelation 55
5. Intestinal Uptake of Free Metal Ions 71
6. Absorption of Metal Ions by Active Transport 103
7. A Secondary Intestinal Pathway for Absorption of Dipeptide-Like Chelates 113
8. Peptide Absorption in the Small Intestine 126
9. Evidence for Two Mineral Absorption Pathways 153
10. The Intestinal Absorption of Dipeptide-Like Chelates as Small Peptides 171
11. Determination of the Locus for Intestinal Absorption of a Dipeptide-Like Chelate 182
12. Mucosal Cell Transport of Dipeptide-Like Chelates as Peptides 197
13. Intracellular Hydrolysis of the Dipeptide-Like Chelate 205
14. Summary and Implications 213

INTESTINAL ABSORPTION OF METAL IONS AND CHELATES

CHAPTER 1

MINERAL NEEDS OF THE BODY

THERE are 92 natural chemical elements on the Periodic Table. At least 50 of them are found in the tissues and fluids of the body. Four of the elements, carbon, oxygen, hydrogen, and nitrogen, contribute about 96% of the total weight of the body,[1] and account for 99% of all of the atoms in the body.[2] The remaining 4% of the body weight and 1% of body atoms are composed of essential and nonessential minerals and mineral contaminants.

Adapting from the criteria established by Cotzias[3] in 1967, a mineral can be classified as essential if (1) it is present in the healthy tissues of the living organism, (2) its concentration in similar animals and man is fairly constant, (3) upon withdrawal from the body, reproducible structural and/or physiological abnormalities result, (4) the addition of the deficient mineral corrects the above-mentioned structural and/or physiological abnormalities, (5) an induced deficiency of a specific mineral is always accompanied by a pertinent specific biochemical change, and (6) the above-mentioned biochemical change can be prevented or corrected when the mineral deficiency is prevented or corrected. It must be remembered, however, that the mere presence of a mineral such as lead or arsenic in the tissues of the animal or man does not prove it is essential. In fact, it may be detrimental to the overall health and performance of the organism.

The essential macrominerals, that is, minerals required in dietary amounts of 100 mg. per day or more, include calcium, phosphorus, sodium, potassium, chlorine, magnesium, and sulfur. Essential microminerals which are needed in quantities of only a few milligrams or micrograms each day are iron, copper, cobalt, zinc, manganese, iodine, molybdenum, selenium, flourine and chromium. Several microminerals that may be essential for life but have not been proven to be definitely needed at present include tin, nickel, silicon, and vanadium. Nonessential contaminants include lead, cadmium, mercury, arsenic, barium, strontium, boron, aluminum, lithium, beryllium, rubidium, as well as others.[1] Figure 1-1 illustrates these concepts as they relate to human beings.[2]

Although the essential microminerals consist of only a small fraction of the total body weight compared to the macrominerals, their value in aiding to extract energy, in growth and maintenance of body tissue, and in assisting in the regulation of body processes[4]

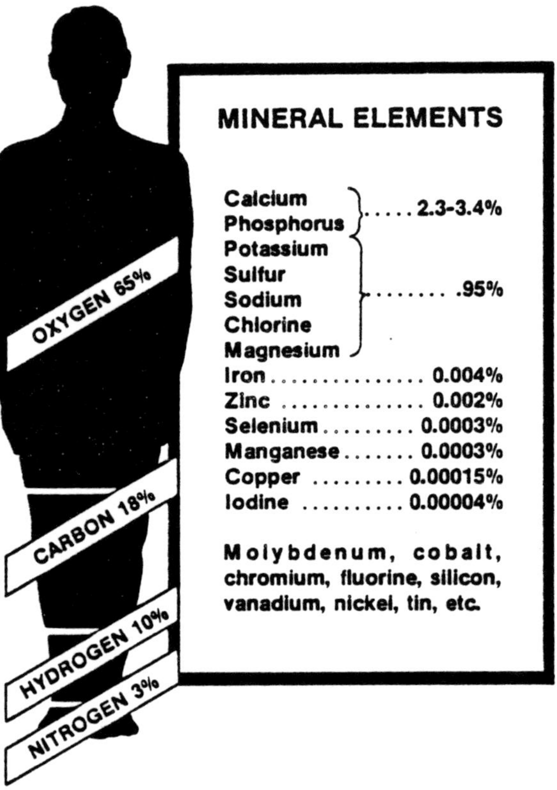

Figure 1-1. Body elemental composition.[2]

cannot be overemphasized. Widdowson and Dickerson have written that, "even small departures from the normal in the mineral composition of the *milieu interieur*, [the environment in which cells live][5] may have profound physiological consequences, but may make no appreciable difference to the composition of the body as a whole."[6]

Underwood has added, "the characteristic concentrations and functional forms of these elements must be maintained within narrow limits if the functional and structural integrity of the tissues is to be safeguarded and the growth, health, and fertility of the animals and man are to remain unimpaired. Continued ingestion of diets, or continued exposure to total environments that are deficient, imbalanced, or excessively high in a particular trace element, induces changes in the functioning forms, activities, or concentrations of that element in the body tissues and fluids so that they fall below or rise above the permissible limits. In these circumstances biochemical defects develop, physiological functions are affected, and structural disorders may arise in ways which differ with different elements, with the degree and duration of the dietary deficiency or toxicity, and with the age, sex, and species of the animal involved."[7]

The science of nutrition, in which minerals play a role, has been defined by the Council on Foods and Nutrition of the American Medical Association as "the science of food, the nutrients and other substances therein, their action, interaction and balance in relation to health and disease and the process by which the organism ingests, digests, absorbs, transports, utilizes and excretes food substances."[8] More simply put, nutrition is the science of nourishing the body properly. This definition applies not only to man but also to other animals.

If the physiological role of food is to provide essential nutrients for normal functioning of the body, then a nutrient is a required substance that the body must obtain externally because it can't manufacture that nutrient or in some cases cannot synthesize it at a sufficient rate to meet the body's needs. Guthrie[4] has categorized these essential nutrients into six basic groups: (1) carbohydrates, (2) fats or lipids, (3) protein, (4) vitamins, (5) water, and (6) minerals. Every one of these nutrients is absolutely essential for the proper functioning of the body. If any one group or member of a group is withheld, death will occur.

By the middle of the 1800s there were clinical indications that minerals, one of the basic groups of nutrients mentioned above, were essential for life. Prior to that time science had learned that carbohydrates, fats, protein and water were not capable of supporting growth of the organism by themselves. Investigators turned to minerals as a clue to the growth-promoting properties of nutrients. In 1857, Claude Bernard[9] published his observations concerning the requirement for iron in respiration. In 1869 an English scientist, Church,[10] reported the biological need for copper in animals. The Americans, Mendel and Bradley,[11] reported in 1905 that zinc was also essential. Then mineral research lapsed into almost complete silence for the next 50 years.

In the meantime, vitamins were discovered. In 1912, Casimir Funk coined the term "vitamine" to describe substances which were necessary for life (vita) and contained nitrogen (amine). With the discovery of the first vitamin, vitamin A, in 1912, research continued to expand in that field until by 1948, fourteen vitamins had been identified.[12]

It was at this point that research returned to minerals since it was recognized that most vitamins required specific minerals in order to function properly. Unfortunately, very little was actually known about this synergistic relationship. As research progressed, it was learned that most of the essential minerals served a variety of functions depending upon their chemical form or combination and their location in the body tissues and fluids. For example, without specific minerals such as calcium and phosphorus the body would have no structural strength.

Minerals are required for a large number of varied purposes in the body. They function as both physical and chemical structural components of body tissue such as calcium and phosphorus in the skeleton, iron in hemoglobin, copper in the metalloenzyme cytochrome oxidase, or cobalt in vitamin B_{12}. Many of these same minerals also function as solutes and electrolytes in solution as free ions in body fluids to help regulate the acid-base balance in the body, to play a role in the osmotic regulation, to aid in the distribution and absorption of body fluids and for secretion, etc.

Information about the functions of minerals in the body has been derived from two basic types of research. The first, which has been applied principally to animals and extrapolated to man, has been to study the consequences of feeding highly purified or specially formulated diets which have a deficiency or an excess of the mineral(s) in question. The second approach has been to study the effects of naturally occurring nutritional diseases due to a mineral deficiency or excess.

"Because trace elements are biologically active at very low concentrations, rigidly controlled experimental conditions are essential. Failure to observe appropriate precautionary measures has often led to erroneous conclusions. Moreover, a given nutrient element may have multiple modes of action and the specific biologic response to this element may occur within a relatively narrow concentration range. At concentrations above this range, the element may actually become inhibitory or deleterious to the organism, a toxic or pharmacologic manifestation resulting from interaction of the element at other sites.

"A further difficulty in the interpretation of many studies results from the frequent occurrence of abnormalities in metal metabolism brought about by *metal ion antagonism* or *conditioned deficiencies* occurring when the normal intake of a nutrient does not meet the needs of an organism because of unusual secondary circumstances or factors. Failure to absorb a metabolite, the inability to synthesize it into a biologically active intermediate and excessive excretion are among the simplest examples of such conditioning factors.

"Another problem in assessing the function of trace metals in biologic processes arises from the fact that their roles are frequently indirect. . . .

"For these reasons and because the determination of concentrations of other metabolites, in tissues, cells, subcellular components and body fluids is a corollary to the elucidation of their role in health and disease, the attention of present-day investigators has centered upon the *distribution* and *function* of metals at the subcellular, enzymatic and molecular levels."[13]

Nutritionists have generally classified the six nutrients including minerals according to their functions.[4] These are seen in Table 1-1.

Energy is derived from carbohydrates, fats, and protein. They are more or less interchangeable in their ability to supply energy depending upon which is more abundant in the diet. Actual energy is obtained only upon oxidation of the energy nutrients. Vitamins and minerals primarily play roles in the oxidation process as enzymatic co-factors. Their roles, and even to a lesser extent the role of protein

Table 1-1
FUNCTION OF NUTRIENTS

SOURCE OF ENERGY	
	Carbohydrate
	Lipid
	Protein
	Minerals
	Vitamins
GROWTH AND MAINTENANCE OF TISSUE	
	Protein
	Lipid
	Minerals
	Vitamins
	Water
REGULATION OF BODY PROCESSES	
	Lipids
	Protein
	Carbohydrates
	Minerals
	Vitamins
	Water

in the enzyme molecule, are indirect rather than directly contributing to the energy supply.[14]

A second function is the growth and maintenance of body tissues. The principal structured materials of the body are water, protein and minerals. Some lipid-containing compounds are also found in the cell membranes. This will be discussed in detail in later chapters. The vitamins do not directly contribute to physical body structure. Nevertheless, growth and maintenance of body tissues is impossible without them. For example, vitamin A does aid in structural requirements of the retina of the eye *per se*, but without it the formation of the retina would not occur. The major role of the vitamins as a whole is in the co-factors of specific enzymes.[14]

The functioning as an enzymatic co-factor or activator is also an equally important role of minerals. For building protein tissues the different proteins that are ingested are not completely interchangeable. They must be digested into amino acids enzymatically and by changes in pH, before they can be restructured into new proteins. Many of the enzymes involved in these processes require specific minerals as activators or co-enzymes.[15]

All of the nutrients—protein, carbohydrates, lipids, minerals, vitamins and water—help regulate body processes. Each nutrient by itself, or more generally in conjunction with other nutrients, performs certain functions that are essential to the normal body metabolism. For example, these include movement of fluids, control of the acid-base balance in the body, coagulation of blood, the activation of enzymes (to carry out these as well as the other two basic functions of nutrients mentioned above), and the maintenance of normal body temperatures, etc.

A more detailed discussion of the function of nutrients is outside the scope of this presentation. Nevertheless, it should be remembered that an individual nutrient, such as a mineral, may fulfill all three of these functions. For example, calcium is absolutely essential for the growth and maintenance of ossa. The amount of calcium required for this bone growth varies with skeletal development, but even at the adult state dietary calcium is constantly required to replace dissolved os for bone maintenance.

Not only is calcium involved in the growth and maintenance of body tissue, but the same metal regulates certain body functions. When a nerve impulse arrives at a neuromuscular junction acetylcholine is released. Acetylcholine is involved in several enzymatic reactions that change the permeability of the nerve cell membranes. When this happens, Na^+ and K^+ move through the membrane creating an electrical impulse. This makes it possible for the nerve impulse to pass to the muscle. Calcium promotes the release of the acetylcholine[16] and the sliding process of muscle contraction.

"The stimulus that produces muscle contraction is an electrical impulse delivered by the motor nerve to the muscle cell. It travels along the membrane of the muscle fiber. The electrical activity increases the permeability of the membranes of the sarcoplasmic reticulum to calcium ions. The ions pour into the cytoplasm, activating the troponin contractile mechanism. Troponin inhibits muscle contraction by blocking the interaction of the protein's action and myosin unless it is combined with calcium.

"After contraction, calcium is quickly removed and returned to the storage sacs in the sarcoplasmic reticulum, and the muscle relaxes. Potassium, sodium, and magnesium are also essential components in the muscle contraction-relaxation processes."[17]

This same mineral that is part of the bone structure and used to regulate body processes is also necessary for the production of energy in the body. Calcium is absolutely necessary to activate the fat-splitting enzyme, pancreatic lipase.[18] "The cells of the pancreas that secrete insulin must have calcium in the intercellular fluid before they can respond to the stimulation from glucose."[19]

Enzymes are extremely important when considering the function of minerals because the major role of the metals in carrying out their three functions is related to their various roles as co-factors or as influences on the enzymes. Through catalytic action, enzymes promote and regulate the multitude of chemical reactions essential for life within the organism.[20]

As an organic catalyst which facilitates specific reactions, enzymes are manufactured by cellular protoplasm and are either secreted into the body fluids or are produced and used within the cytoplasm of the body cells themselves. It is estimated that the average animal cell contains several thousand different enzymes. Enzymes are proteins (polymers) composed of long chains of relatively low molecular weight amino acids. These amino acids contain both the alkaline amino (NH_2) group and the acidic carboxyl (COOH) group.[20] This type of structure allows other compounds to be linked to these chains of amino acids through the process of chelation,[21] which will be discussed in detail in Chapter 4.

Many enzymes are more than the simple proteins referred to above. Through chelation these specific complex proteins are linked to nonproteinaceous substances called prosthetic groups containing either a vitamin or a mineral. When the prosthetic group is part of the enzyme, the enzyme is called a metalloenzyme. When the prosthetic group is readily removed from the enzyme, it is called a cofactor.[22] In this case, through chelation or complexing, a simple ion is attached to the protein portion and activates the enzyme. When the ion is removed from the enzyme, all of its catalytic ability is lost.

In Eichhorn's discussion of enzymes he indicates that if the metal is weakly bonded to the protein, the ion may participate in bond cleavage reactions by altering the enzyme's structure through pulling away the negative charge from the substrate and then attaching the enzyme to the substrate. If the metal is strongly bound to the protein molecule of the enzyme, then the ion participates in an electron transfer type of reaction. Basically, the catalytic activity of the enzyme is determined by how the protein is attached to the prosthetic group and the electronic configuration of the metal in the prosthetic group.[21]

The cell is basically a metal deficient system in which all of the ligands, including those forming enzymes, are in competition for a limited number of metal ions.[23]

"It is because they are essential components of certain enzyme systems that . . . [minerals] play such an important role in biological processes, that deficiencies or excesses of them can seriously alter the enzymatic equilibrium of cells."[24]

In summary, it has been seen that man and animals are autonomously functioning complex units whose basic structure is composed of carbon, hydrogen, oxygen, and nitrogen. In addition to these basic elements, living cells also require the macrominerals sulfur, phosphorus, sodium, potassium, chlorine, calcium and magnesium. They are basic to the composition of the body. Finally, to complete the life processes, trace minerals are required for many vital functions. The great majority of the trace elements serve as key components of enzymatic systems or in proteins with other vital roles. Although the macrominerals' chief role is structural, they too can function in enzymes or in proteins. Many clinical and pathological disorders arise as a consequence of mineral deficiencies and/or excesses. This illustrates the numerous roles and interrelationships of minerals in nutrition, in health, and in disease and underline the importance of optimum mineral nutrition.

REFERENCES

1. Anderson, C., "Minerals" in Schneider, H., *et al.*, ed., *Nutritional Support of Medical Practice* (Hagerstown: Harper & Row), 57-8, 1977.
2. Briggs, G. and Calloway, D., *Nutrition and Physical Fitness* (Philadelphia: W. B. Saunders Co.), 236-7, 1979.
3. Cotzias, G., Hemphill, D., eds., *Proc. 1st Ann. Conf. Trace Substances Environmental Health* (Columbia: U. of Missouri), 5, 1967.

4. Guthrie, H., *Introductory Nutrition* (St. Louis: C. V. Mosby Co.), 11, 1975.
5. Bernard, C., *Lecons sur les Phenomenes de la vie Communs aux Animaux et aux Vegetaux* (Paris: Bailliere), 121, 1818.
6. Widdowson, E. and Dickerson, J., "Chemical Composition of the Body," in Comar, C. and Bronner, F., eds., *Mineral Metabolism* (New York: Academic Press,), 2:57, 1964.
7. Underwood, E., *Trace Elements in Human and Animal Nutrition* (New York: Academic Press), 7, 1977.
8. Guthrie, op. cit., 3.
9. Bernard, C., *Lecons sur les Effects des Substances Toniques et Medicamenteuses* (Paris: Bailliere), 1857.
10. Church, A., Phil. Trans. Roy. Soc. London, 159:627, 1869.
11. Mendel, L. and Bradley, H., Am. S. Physiol., 14:313, 1905.
12. Guthrie, op. cit., 5
13. Fein, J., "Nutrition in Diseases of Gastrointestinal Tract," in Goodhart, R. and Shils, M., eds., *Modern Nutrition in Health and Disease* (Philadelphia: Lea & Febiger), 910, 1980.
14. Wilson, E., *Principles of Nutrition* (New York: John Wiley & Sons), 4, 1979.
15. Schutte, K., *The Biology of the Trace Elements* (Philadelphia: J. B. Lippincott Co.), 17, 1964.
16. White, A., *et al.*, *Principles of Biochemistry* (New York: McGraw Hill Book Co.), 961, 1973.
17. Wilson, op. cit., 266.
18. Gallagher, C., *Nutritional Factors and Enzymological Disturbances in Animals* (Philadelphia: J. B. Lippincott Co.), 38, 1964.
19. Guthrie, op. cit., 115.
20. Schutte, op. cit., 14.
21. Eichhorn, G., "The Role of Metal Ions in Enzyme Systems," in Seven, M., ed., *Metal Binding in Medicine* (Philadelphia: J. B. Lippincott Co.), 19, 1960.
22. Gallagher, op. cit., 22.
23. Williams, R., "Coordination, Chelation and Catalysis," in Boyer, P., *et al.*, eds., *The Enzymes* (New York: Academic Press), V1, 412, 1951.
24. Schutte, op. cit., 15.

CHAPTER 2

FACTORS AFFECTING MINERAL ABSORPTION FROM THE INTESTINE

AS indicated in the previous chapter, structure, the activation of intracellular and extracellular enzymes, regulation of critical pH levels in body fluids necessary for the control of metabolic reactions and the osmotic balance between the cell and its environment, are all dependent upon minerals. A deficiency of any one of the essential minerals mentioned in Chapter 1 can, and often does, result in severe problems to the organism.

Pike and Brown[1] in their "Summary of Nutrient Metabolism" have presented a simplified basis for metabolic experiments which is reproduced in Figure 2-1. The factors most commonly measured are those indicated in bold type.

While some of these factors are valid measurements of mineral metabolism, no single parameter is sufficient for accuracy. It is only through biochemical evaluations of nutrients, nutrient metabolites, or related functional and structural components that a picture begins to emerge concerning the numerous influences on mineral absorption and metabolism. Predictability of absorption is possible only within wide limits. The purpose of this chapter is to examine some of these influences that affect mineral uptake.

Minerals are frequently interrelated to each other. A deficiency of one may affect the efficient functioning of another. Liebig, who formulated the "Law of the Minimum," stated that the nutrient that is in the relative minimum determines the overall performance of the organism.[2] In this discussion, then, the level of mineral absorp-

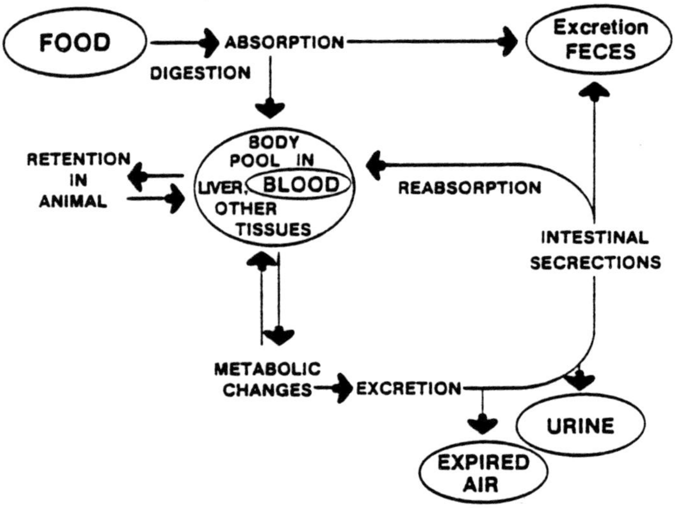

Figure 2-1. Summary of nutrient metabolism.[1]

tion and metabolism is the nutrient under consideration.

To illustrate, zinc is involved in the synthesis of protein through its activation of the enzyme DNA polymerase. Starting with a pool of nucleoside triphosphates, DNA polymerase synthesizes it into deoxyribonucleic acid (DNA). DNA polymerase is a metalloenzyme which requires zinc for its activation.[3] Thus, zinc plays a specific role in protein synthesis. A deficiency of this metallic nutrient will as a consequence affect protein synthesis.

A zinc deficiency also causes a sharp decrease in the cellular production levels of RNA as well as the ribosome content of the cells. This reduction leads to an inhibition in protein formation while glucose and non-protein nitrogen increase.[57]

The B vitamin, niacin, is also involved in protein synthesis. It functions in conjunction with zinc as a coenzyme in DNA polymerase.[4] It is theorized that the niacin picks up the zinc and places it in the DNA polymerase enzyme. Thus, without the niacin in conjunction with the zinc, protein synthesis is also impossible. But Harper emphasizes that niacin deficiencies seldom occur.[4]

Besides zinc and niacin, protein synthesis also requires amino acids for restructuring new peptides into new polymeric chains. The DNA supplies the new protein template information through RNA, the sequence of which dictates the correct production of these protein chains.[5]

In addition to the amino acids, the niacin, and the zinc, protein synthesis requires other substances as well.[5] The point is that if any one of them is deficient, then according to Liebig's Law, the deficient substance will affect the growth of the organism.[2] To illustrate, in research done by Hsu,[6] the effects of only one nutrient deficiency, zinc, on the levels of free amino acids in urine, plasma, and skin were studied. When there was a zinc deficiency, there were significantly greater concentrations of free amino acids in the urine. This suggested that due to a zinc deficiency, there was an inability for the body to metabolize all of the available amino acids consumed—thus they were excreted into the urine as waste.

The fact that these same zinc-deficient test animals also had lower values of many amino acids in the skin extracts, also seemed to confirm this zinc-amino acid relationship observation. Hsu concluded that protein synthesis in the skin was related to the quantity of zinc available for activation of enzymes to restructure the large quantity of amino acids into new polypeptide chains.[6] In fact, Praske and Plocke observed that a zinc deficiency resulted in unstable cytoplasmic ribosomes which can be related to a reduction in RNA synthesis is also related to the above.[57]

This example effectively illustrates the "Law of the Minimum."[2] The level of zinc in the body determines the overall ability of the cells to produce new protein for growth.

The late Professor Eric Underwood stated, "Metabolic interactions among trace elements are so potent and so diverse that no consideration of the current status of nutrition would be reasonable without some account of their nutritional implications . . . trace metal interactions are more frequent among elements that share common chemical parameters and compete for common metabolic sites."[7]

All minerals interact with and influence each other, and a balance between them must be maintained if health and nutrient efficiency are to be achieved. As a general rule, those minerals that are involved in many metabolic processes are more likely to be interrelated, one to another, than those involved in a single function.[8]

One of the first valid studies on mineral interactions was published in 1954 by Dick.[9] He reported that there was a three-way interaction between copper, molybdenum and sulfur as a sulfate. Since

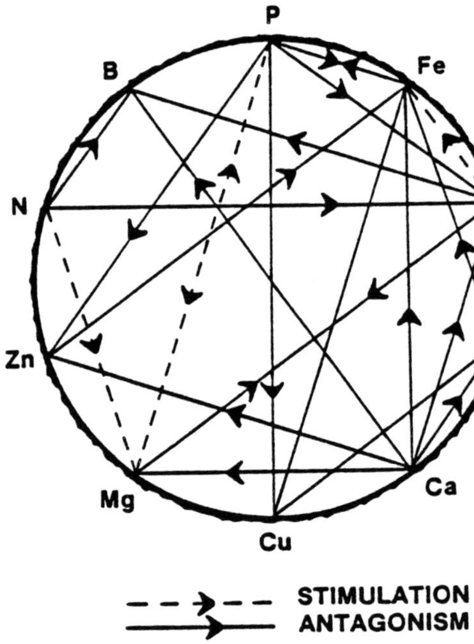

Figure 2-2. Mulder's mineral interactions in plants. Solid lines indicate antagonism and dotted lines indicate stimulation.[2,10]

that first study, other mineral relationships have been investigated and established as valid interactions.

Similar studies were being conducted in plants at about the same period of time. In 1953, Mulder first showed the complexity of mineral reactions even in plants with his Interaction Chart.[2,10] This is seen in Figure 2-2.

Using this mineral interaction concept for only one application, that of intestinal absorption of ions from ingested soluble inorganic mineral salts that are competing for carrier molecules[11] and/or intestinal absorption sites,[12,13,14] Dyer expanded Mulder's Interaction Chart and applied it to animals. Even so, he only showed the antagonism between the various mineral nutrients when ingested as salts.[15] This is seen in Figure 2-3. The arrows indicate antagonism between minerals as they compete for absorption sites and/or carrier molecules.

Suttle[16] has grouped these mineral interactions shown in Figure 2-3 into six categories according to the type of mechanism involved

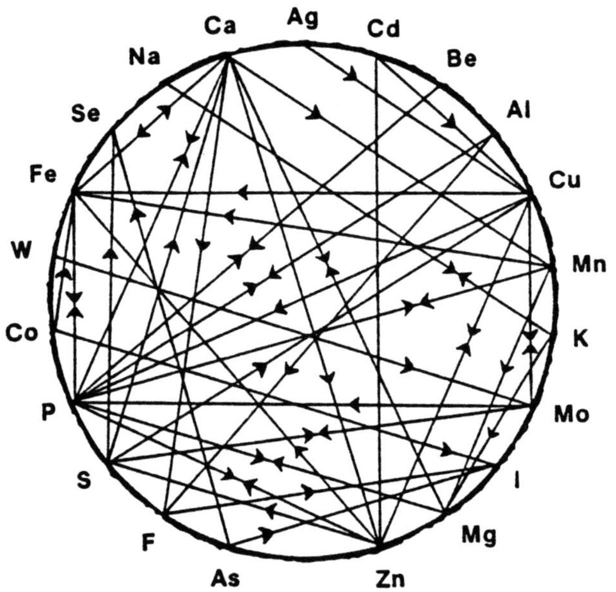

Figure 2-3. Mineral antagonism from salts in the intestine.[15]

in causing the interaction which in turn influences absorption of each mineral.

The first category involves the formation of insoluble complexes. The primary action often involves only one mineral plus a precipitating ligand. The interaction occurs when minerals compete for the same anionic ligand. The ligand may be inorganic, such as a phosphate,[17,18] or organic such as a phytate.[19]

The complexing with a precipitating ligand can occur in the tissues wherein a mineral can displace another mineral from its position in the tissue. The displaced mineral is then free to enter into a reaction with a precipitating ligand.[20,21,22]

More appropriate to this discussion are the interactions occurring in either the diet[23,24] and/or the digestive tract.[25,26] Again there is competition among the minerals for the precipitating ligand. The competition depends upon the concentrations of the minerals involved, the relative strengths of the association constants of the minerals, and the solubility of the product.[16]

Precipitating substances derived from the diet are a major factor affecting intestinal absorption of metal. Once the mineral salt is ionized in preparation for membrane transport of the cation, it is by its very physical nature unstable. In the ionic state it is highly susceptible to sequestering by phytic acid from cereals and other plant seeds. The phytic acid forms very stable complexes with mineral ions rendering them unavailable for intestinal uptake because the first step in mineral absorption requires that the mineral remain in the ionic state.[19] As the phytic acid content of the diet increases, the intestinal absorption of certain metals, specifically zinc and calcium and presumably others, decreases.[27] The reduction in mineral phytate absorption may be due to the lack of ionization of the mineral complex in the stable phytate form. If precipitation were to occur, there would be no inducement for carrier proteins[28] located on the intestinal cell membrane to bind themselves to the metal ion prior to membrane transport[11] because the metal is not in ionic form.

The same general principle holds true for other mineral precipitating substances found in the diet such as oxalates, and probably more importantly from a dietary point of view, phosphates.[17,18] Calcium, magnesium, zinc, iron, aluminum and beryllium all react with dietary phosphates to form insoluble precipitates.[29] Most phosphates are slightly soluble in water or acid solutions. However, as will be shown in Chapter 3, the intestine tends to become alkaline[30] which reduces the solubility of the phosphates when introduced into that environment. When the mineral is trapped within an insoluble compound, there is little likelihood of significant intestinal absorption since the precipitated salt is generally stable, creating no attraction to induce integral carrier proteins in their lipid environment in the membranes of the mucosal cells to bind the ion and move it across the membrane.

The second group of mechanisms that influences intestinal absorption of minerals involves the competition of chemically similar ions for the same transport carriers.[31] The gut is an important site for these interactions. For example, carrier proteins[28] of small molecular weight can be found in the duodenum. These carrier proteins are subject to competition among the ions for their binding and consequential absorption. The competition can be between trace elements, macrominerals or between micro- and macrominerals.[32]

The investigations[33] of El-Shobaki and Rummel illustrate this interaction. They found that copper is preferentially bound to transferrin, the protein transport molecule in the mucosa, when competing with iron. Normally this transport mechanism is not completely saturated, so there are adequate binding sites for both the iron and the copper. Nevertheless, when copper and iron are administered to an excess, iron absorption is inhibited because of the preferential binding of copper to the transferrin. This displacement is based on the ability of the organic molecule to change the electromotive potential of the iron[34]. In other words, some metals will be bound to the carrier protein before others, and a displacement can occur if a stronger metal requires the binding site held by a weaker metal.

The third mechanism that may affect mineral uptake is not directly related to intestinal absorption, although there is some evidence that a similar mechanism may possibly exist in the gut. It involves the synthesizing of metal-binding proteins by the body as a specific reaction to heavy metal loading.[35,36] The proteins are not specific for the toxic mineral *per se*, and they may bind other minerals, thus participating in trace mineral interactions. The full significance of these metalloproteins is yet to be elucidated.

A change in the metal component of a metalloenzyme is the fourth mechanism involved in mineral absorption. As previously alluded to, a metal that activates a specific enzyme can be replaced by another metal that may either block or accelerate that particular enzymatic activity. For example, the carboxypeptidase enzyme is activated by zinc. The zinc can be replaced by cobalt, which will cause a decrease in peptidase activity.[37]

Most mineral competition obviously occurs in the plasma and tissues instead of the gut, because the majority of the enzymatic activity within the body occurs in the tissues and plasma. Nevertheless, digestive enzyme activity does influence mineral absorption. The competition of different minerals for those enzymes therefore influences their overall absorption.

The fifth mechanism involving mineral interactions is related to the second mechanism, that of metabolic pathways. More specifically, it involves the transport and excretion of the minerals. Some of these transport mechanisms are in the intestinal cell membrane, while others are in the mucosal cells themselves. From available data from the plasma and extrapolated to the mucosa it appears that only specific interrelationships are involved as they relate to specific transport mechanisms.[38]

The last mechanism has the least significance to mineral nutrition, but nevertheless is important in the overall view of mineral interactions. In each of the above, the mineral interaction has been considered separately, but many interactions are precipitated and/or affected by other interactions occurring concurrently. To illustrate, the precipitating of a mineral with an insoluble ligand will interfere with the ability of that specific metal to enter into other enzymatic interactions[16] which in turn can affect the uptake of still other minerals which were not affected by the precipitating ligands directly.

Suttle adds that the major problem of these mineral interactions is that the chemical form in which the mineral is presented to the digesta, as well as the form in which it is absorbed from the gut, affects its susceptibility to interaction. Thus, true physiological conditions are difficult to define.[16]

Not only do minerals interfere with the absorption of other minerals as indicated above, but other nutrients, such as vitamins, can have equally significant positive or negative influence on mineral absorption. Vitamin D is one such example. Its role in the regulation of intestinal absorption of ionic calcium has been conclusively proven.[35] The vitamin D must be altered metabolically to regulate the calcium absorption.[36] The change commences in the liver, followed by additional changes in the kidney to produce a hormone, 1,25 dihydroxyvitamin D_3. This hormone controls the absorption of the calcium ion from the intestine by stimulating the active transport of calcium against an electrochemical potential gradient.[37] The hormone 1,25 dihydroxyvitamin D_3 is also involved in supplying genetic instruction for the production of the carrier protein across the mucosal cell to the serosa [38] as seen in Figure 2-4.[56] A deficiency of vitamin D will interfere in the absorption of ionic calcium.

A second vitamin, ascorbic acid, when consumed in orange juice will enhance the absorption of non-heme from 3.7 percent to 10.4 percent.[39] Conrad and Schade[40] have demonstrated a probable role for the ionic iron absorption enhancement role of vitamin C. This is seen in Figure 2-5.

In the absence of reducing agents, iron exists in the ferric form. This ferric iron has a valence of +3. As a result, when it is in an aqueous solution such as in the gut, the metal ions are bound to each other through water bridges. As the pH in the intestine rises,[30] more hydroxyl ions become available to combine with the ferric ions.

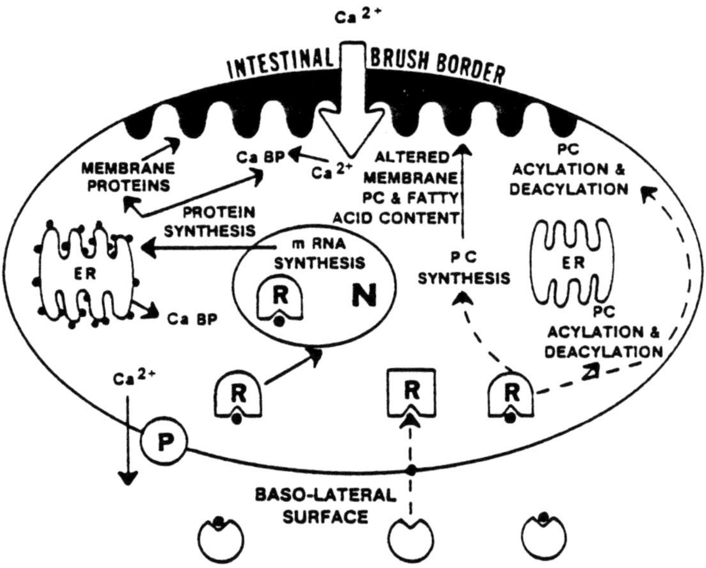

Figure 2-4. The effect of $1,25(OH)_2D(\bullet)$ on metabolism of calcium. $1,25(OH)_2D$ bound to DBP in tissue fluid enters the cell where it binds to a receptor (R). This stimulates various metabolic events resulting in an increased uptake of Ca^{2+} from the intestinal lumen by possibly increasing Ca permeability at brush border. The transfer of the calcium across the mucosal cell by either vesicles or mitochondria has not been clearly established. Ca^{2+} is eventually actively transported out of the cell by an enzyme (P) located in the baso-lateral membrane. Dotted arrows represent pathways about which very little is known.

These iron hydroxyl molecules form precipitates that are inaccessible to the mucosal cells because the iron is no longer in an ionic state which is necessary for its mucosal uptake.

The addition of the ascorbic acid to the aqueous acid solution of the stomach and the upper part of the duodenum causes a displacement of hydrogen ions from the resulting iron ascorbate chelate. As seen in Figure 2-5, this iron chelate has the ability to remain in the solution over a pH range of between 2 and 11. (It should be stressed that the reaction is initiated optimally at an acid pH because the ferric chloride as well as the ascorbate are both insoluble at an alkaline pH). Thus, the presence or absence of vitamin C will affect the absorption of this particular form of mineral.

Other nutrients, such as fats, can also affect the quantity of a given mineral's absorption in the intestine. The absorption of cal-

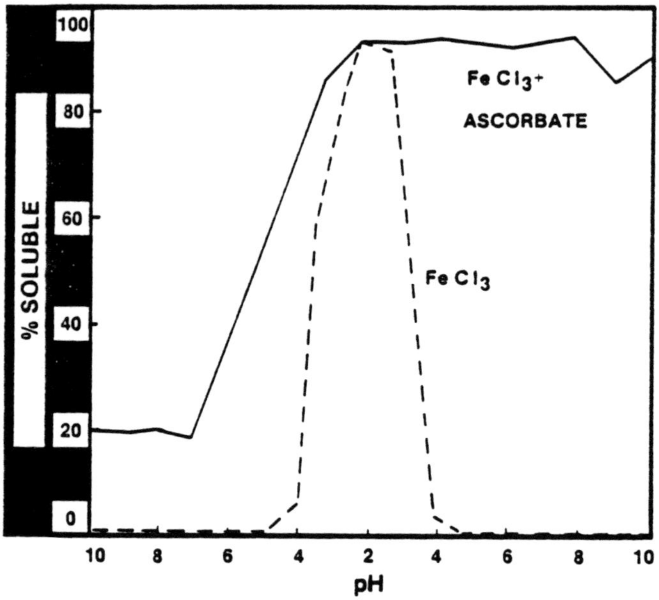

Figure 2-5. The effects of pH on the solubility of $FeCl_3$ with and without ascorbate. Both $FeCl_3$ (dotted line) and $FeCl_3$ (solid line) to which an equimolar quantity of ascorbic acid had been added were relatively insoluble in NaOH. As the pH was reduced by the addition of HCl, both became soluble. $FeCl_3$ was again precipitated by realkalization with NaOH, while $FeCl_3$ with ascorbic acid remained soluble.[40]

cium can be depressed in the presence of high levels of fat, particularly those with higher melting points.[41] Some unsaturated fatty acids also form calcium salts which are poorly absorbed. Thus, the degree to which the fats interfere with intestinal absorption depends on the type of fat in the diet.[42] Presumably, if calcium can be bound by the unsaturated fatty acids, other bivalent metal ions can be tied up also. The reason calcium is measurable is because it is a macromineral in the diet. The U.S. recommended daily allowance (U.S.R.D.A.) for calcium in man is 1000 milligrams per day. On the other hand, a trace element such as iron, has a U.S.R.D.A. of only 18 mg. Thus, with the smaller amount in the diet, the effect of the fat on its absorption is probably negligible. Indeed, Briggs and Calloway have reported that even with calcium, the amount of insoluble soaps formed is not usually a significant factor in absorption.

One last dietary item needs to be mentioned. Fiber has been found to be responsible for preventing the absorption of many, and perhaps all, minerals. In a balance study, two adults' diets were changed from one low in fiber content to one of high fiber content. Calcium, magnesium, zinc, and phosphorus absorption were decreased. The loss of each metal in the feces correlated very closely to the increased fecal dry weight, which in turn was directly proportioned to the fecal fiber excretion.[44] The reduction of metal absorption was mainly due to its absorption in the non-digestive cellular fibers.

Certain forms of mineral salts require a very low pH to solubilize. They will remain insoluble in an alkaline or slightly acid environment. In an alkaline pH, such as occurs in the intestine below the bile duct, there appears to be little attraction for the carrier protein to be attached to the mineral and carry it across the membrane into the mucosal cells. A very small amount of these minerals may be absorbed into the cells through diffusion or through some natural chelation consequential to pH changes if the ions have remained soluble, but the nutritional significance of that might be questionable.

The above examples vividly illustrate the necessity of creating a metal ion state if absorption of the mineral from the lumen into the mucosal cell is to occur. Different regions of the small intestine vary in their capacities to absorb the mineral. There is a decreasing absorption rate as the ion traverses the intestine[45] as shown in Figure 2-6.

The availability of the metal ion for mucosal absorption is generally dependent upon the pH of the intestinal lumen. Many metals, with the exception of the alkali metals, have a tendency to form insoluble salts with either anions in the diet or of physiological origin in the higher pH range of the small intestine. In general, the more alkaline the lumen, the lower the rate of absorption of most minerals.[30]

As has been shown, the transport of a given mineral through the mucosal cell membrane is due to the affinity of the metal for the binding carrier protein or in the cell membrane. The total capacity of the cell membrane to bind or retain the metal is obviously a major consideration in determining the total quantity absorbed.

Any factor that inactivates the membrane binding sites or alters the binding characteristics of the ion results in lower transport across

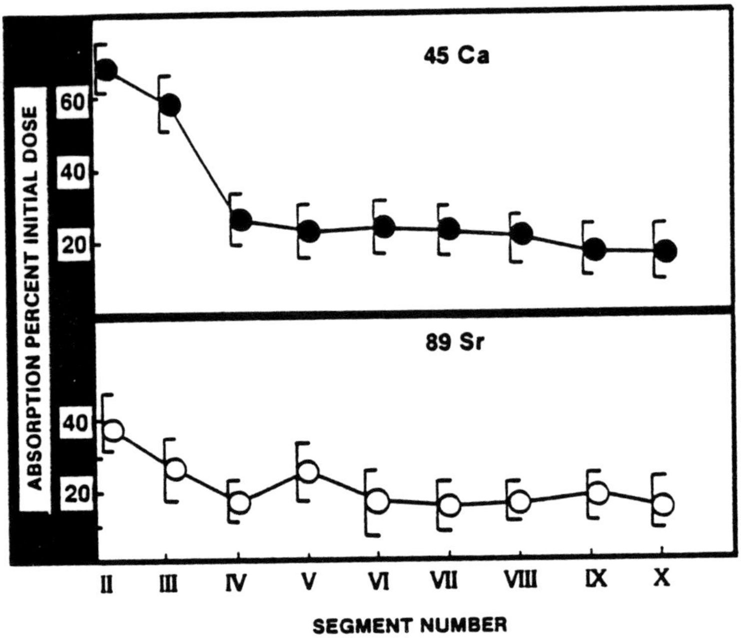

Figure 2-6. Absorption of Ca and Sr in the intestine.[45]

the membrane. Thus, ion dependent absorption presents a real challenge: How much of a given mineral is actually absorbed from the lumen? There is no general agreement.

To illustrate, Guthrie[46] has reported that intestinal absorption of calcium ranges between 20% and 30% of the ingested amount under normal circumstances. Briggs and Calloway[47] disagree. They report 20% to 40% absorption as being normal. Guthrie counters with 30% to 50% as being optimum and 10% absorption frequent.[48]

Underwood has reported intestinal absorption of zinc ranges between 4.8% and 13.4%.[58] Guthrie claimed the absorption is closer to 20%, with 30% being the maximum.[59]

As far as magnesium absorption is concerned Guthrie has stated it ranges between 25% and 75% of the administered dose with 43% being the mean.[60]

Data regarding the absorption of other minerals are equally confusing. Absorption of iron ranges between 2% and 10%, according to Guthrie.[46] Underwood states absorption is between 2% and 20%, with the narrow range of 2% to 4% being normal.[49] Briggs and Calloway state maximum absorption in cases of iron deficiency anemia is 40% of the ingested amount.[50] Guthrie says it is 60%.[48]

Values for manganese absorption range between 3% and 4% all the way up to 20%, depending upon the authority consulted.[51,52] Guthrie feels 30% absorption for copper is normal, whereas Underwood claims it is 10%. Guthrie's values are higher, claiming absorption ranges between 15% and 97%.[53] And so the controversy goes on.

Even in controlled *in vitro* intestinal studies, there is no general agreement on absorption of the various metals.[34,35] In these studies, adult male Sprague-Dawley rats were sacrificed on the day of experimentation by cervical dislocation. Their intestines were quickly removed, washed three times in Krebs-Ringer's Bicarbonate Buffer (ph 7.4) and the portion starting after the duodenum cut into two centimeter segments that had been severed longitudinally along the mesenteric line. Randomized samples from several rats were preincubated for one minute in additional Krebs-Ringer's Bicarbonate Buffer (ph 7.4) while 0_2 (95.5%) and CO_2 were bubbled through fritted glass tubes. Triplicate samples were then incubated two additional minutes in glass beakers containing the various mineral salts in solution through which the same gas mixture was bubbled. Upon removal from incubation, the intestinal segments were washed six times, dried, weighed and assayed for metal uptake by atomic absorption spectrophotometry. Table 2-1 presents the mean results from some of these tests.[55]

Inasmuch as the rats had received a well-fortified laboratory diet (Purina Rat Chow), the minerals were presumably at optimum levels in their bodies. The mineral assimilation into the intestinal cells was not a forced condition brought about by an acute physiological need as may have been the case had the animals been mineral deficient. The point to remember is that even though there was equal opportunity to absorb similar quantities of the ionic metals in solution and that all the salts were dissolved, actual absorption varied according to the salt used.

These data further tend to emphasize the predictive problem with mineral absorption. There is neither consistency nor general agreement as to what constitutes normal mineral absorption from the intestines. The multitude of external environmental factors in the gut can, and often do, interfere with mineral absorption, making

Table 2-1.

INTESTINAL ABSORPTION OF MINERALS				
	Control	Carbonate	Sulfate	Oxide
Copper	Trace	6	8	11
Magnesium	7	51	36	23
Iron	23	53	78	61
Zinc	14	87	84	66
(Data are expressed as ppm)				

precise predictions impossible. Based on the very critical body requirements for many of these minerals needed to insure optimum performance, greater consistency is essential. The purposes of the study to follow is to review the evidence for a secondary mineral absorption pathway that functions outside the generally accepted models for ionic mineral absorption, a pathway which may be more effective in the intestinal uptake and transport of those minerals.

REFERENCES

1. Pike, R. and Brown, M., *Nutrition: Integrated Approach* (New York: John Wiley & Sons), 819, 1975.
2. Schutte, K., *The Biology of Trace Elements* (Philadelphia: J. B. Lippincott Co.), 43, 1964.
3. O'Dell, B., "Introduction", in Pories, W., *et al* ed., *Clinical Application of Zinc Metabolism* (Springfield: Charles C Thomas), 5, 1974.
4. Harper, H., *Review of Physiological Chemistry* (Los Altos: Lange Medical Publications), 99 & 169, 1971.
5. Ibid, 65.
6. Hsu, J., "Zinc Deficiency and Alterations of Free Amino Acid Levels in Plasma, Urine and Skin Extract," in Brewer, G. and Prasad, A., eds., *Zinc Metabolism: Current Aspects in Health and Disease* (New York: Alan R. Liss, Inc.), 73, 1977.
7. Underwood, E., "The Current Status of Trace Element Nutrition: An Overview," a paper presented at International Minerals Conference, St. Peters-

burg Beach, Fla., January 17, 1978.

8. Dyer, I., "Mineral Requirements," in Hofez, E. and Dyer, I., eds., *Animal Growth and Nutrition* (Philadelphia: Lea & Febiger), 312, 1969.
9. Dick, A., "Studies on the Assimilation and Storage of Copper in Sheep," Aust. J. Agri. Res., 5:511, 1954.
10. Mulder, D., "Les Elements Mineurs en Culture Furitiere," 1° Convengno Nazeonale de Frutticoltura Montana di Saint Vincent, 188, 1953.
11. Sahagian, B., *et al.*, "Effects of Inhibitors and Accelerators on Intestinal Absorption of Divalent Trace Metals," in Skoryna, S. and Waldron-Edward, D., eds., *Intestinal Absorption of Metal Ions, Trace Elements and Radionuclides* (Oxford: Pergamon Press), 321, 1971
12. Mountcastle, V., *Medical Physiology* (St. Louis: C.V. Mosby Co.), 2, 1146, 1974.
13. MacInnis, A., *et al.*, "Specificity of Amino Acid Transport in the Tapeworm Hymenolepsis Diminuta and its Rat Host," Rice University Studies, 62:183, 1977.
14. Witt, W., *Biology of the Cell* (Philadelphia: W.B. Saunders Co.), 407-8, 1977.
15. Dyer, op. cit., 313.
16. Suttle, N., "Trace Element Interactions in Animals," in Nicholas, D., and Egan, A., eds., *Trace Elements in Soil Plant Animal Systems* (New York: Academic Press, Inc.), 271, 1975.
17. Williams, S., *Nutrition and Diet Therapy* (St. Louis: C.V. Mosby Co.), 137, 1977.
18. Cabell, C. and Earle, I., "Additive Effect of Calcium and Phosphorus on Utilization of Dietary Zinc," J. Animal Sci., 24:800, 1965.
19. Vohra, P., *et al.*, "Phytic-Acid Metal Complexes," Proc. Soc. Exp. Biol. Med., 120:447, 1965.
20. Mills, C., *et al.*, "Metabolic Effects of Molybdenum Toxicity in the Rat," J. Nutr., 65:129, 1958.
21. Rish, M., "The Geochemical Ecology of Organisms in Deficiency and Excess of Copper," in Mills, C., ed., *Trace Element Metabolism in Animals* (Edinburgh: E & S Livingstone), 452, 1970.
22. Dowdy, R. and Matrone, G., "Copper-Molybdenum Interaction in Sheep and Chicks," J. Nutr., 95:191, 1968.
23. Vohra, P. and Kratzer, F., "Influence of Various Phosphates and Other Complexing Agents on the Availability of Zinc for Turkey Poults," J. Nutr., 89:106, 1966.
24. Waddell, D. and Sell, "Effects of Dietary Calcium and Phosphorus on the Utilization of Dietary Iron by the Chick," J. Poultry Sci., 43:1249, 1964.
25. Taylor, T., "The Availability of the Calcium and Phosphorus of Plant Materials for Animals," Proc. Nutr. Soc., 24:105, 1965.
26. Davis, N. and Nightingale, R., "Effect of Phytate on Zinc Absorption and Fecal Zinc Excretion and Carcass Retention of Zinc, Iron, Copper and Manganese," Proc. Nutr. Soc., 34:8A, 1975.
27. Oberleas, D., *et al.*, "The Availability of Zinc from Foodstuffs," in Prasad, A., ed., *Zinc Metabolism* (Springfield: Charles C Thomas), 225, 1966.

28. Witt, op. cit., 433.
29. Thompson, D., *Trace Elements in Animal Nutrition* (Skokie: IMC Corp.), 26, 1970.
30. Waldron-Edward, D., "Effects of pH and Counter-Ion on Absorption of Metal Ions," in Skoryna, S. and Waldron-Edward, D., eds., *Intestinal Absorption of Metal Ions, Trace Elements, and Radionuclides* (Oxford: Pergamon Press, 381, 1971.
31. Hill, C. and Matrone, G.: "Chemical Parameters in the Study of *in vivo* and *in vitro* Interactions of Transition Elements," Fed. Proc., 29:1474, 1970.
32. Starcher, B., "Studies on the Mechanism of Copper Absorption in the Chick," J. Nutr., 97:321:1969.
33. El-Shobaki, F. and Rummel, W., "Binding of Copper to Mucosal Transferrin and Inhibition of Intestinal Iron Absorption in Rats," Res. Exp. Med., 174:187, 1979.
34. Ashmead, H., "Tissue Transport of Organic Trace Minerals," J. Appl. Nutr., 22:42, 1970.
35. Templin, V. and Steenback, H., "Vitamin D and the Conservation of Calcium in the Adult," J. Biol. Chem., 100:209, 1933.
36. Blunt, J. and DeLuca, H., "Biologically Active Metabolites of Vitamin D" in DeLuca, H., ed., *The Fat Soluble Vitamins* (New York: Plenum Press), 69, 1978.
37. DeLuca, H., 'Vitamin D and Calcium Transport," Ann. N.Y. Acad. Sci., 307:356, 1978.
38. DeLuca, H., "The Control of Calcium and Phosphorus Metabolism by the Vitamin D Endocrine System," in Levander, O. and Cheng, L., eds., *Micronutrient Interactions: Vitamins, Minerals, and Hazardous Elements* (New York: New York Academy of Science), 12, 1980.
39. Callender, S., *et al.*, "Eggs and Iron Absorption," Br. J. Nutr., 19:657, 1970.
40. Conrad, M. and Schade, S., "Ascorbic Acid Chelates in Iron Absorption: A Role for Hydrochloric Acid and Bile," Gastroenterology, 55:35, 1968.
41. Nicolaysen, R., *et al.*, "Physiology of Calcium Metabolism," Physiol., Rev., 33:424, 1953.
42. Fleischman, A.I., *et al.*, "Effects of Dietary Calcium Upon Lipid Metabolism in Mature Male Rats Fed Beef Tallow," J. Nutr., 88:255, 1966.
43. Briggs, G. and Calloway, D., *Nutrition and Physical Fitness* (Philadelphia: W.B. Saunders Co.), 252, 1979.
44. Reinhold, J., *et al.*, "Decreased Absorption of Calcium, Magnesium, Zinc, and Phosphorus by Humans Due to Increased Fiber and Phosphorus Consumption as Wheat Bread," J. Nutr,, 106:493, 1976.
45. Waldron-Edward, D., *et al.*, "Effects of the Counterion and pH on Intestinal Absorption of Calcium and Strontium," Proc. Soc. Exp. Biol. Med., 123:532, 1966.
46. Guthrie, H., *Introductory Nutrition* (St. Louis: C.V. Mosby Co.), 116, 1975.
47. Briggs, op. cit., 236.
48. Guthrie, op. cit., 148.
49. Underwood, E., *Trace Elements in Human and Animal Nutrition* (New York Aca-

demic Press), 27, 1977.
50. Briggs, op. cit., 227.
51. Underwood, op. cit., 174.
52. Briggs, op. cit., 292.
53. Guthrie, op. cit., 70.
54. Underwood, op. cit., 70.
55. Graff, D., *et al.*, "Absorption of Minerals Compared with Chelates Made from Various Protein Sources into Rat Jejunal Slices *in vitro*," Address to Utah Academy of Arts, Letters, and Sciences, Salt Lake City, April 1970.
56. Holmes, R. and Kummerow, F., "The Relationship of Adequate and Excessive Intake of Vitamin D to Health and Disease," J. Amer. Coll. Nutr., 2:183, 1983
57. Praske, J. and Plocke, D., "A Role for Zinc in the Structural Integrity of the Cytoplasmic Ribosomes of *Eugelena Gracilis*," Plant Physiol, 48:150, 1971.
58. Underwood, op. cit., 205.
59. Guthrie, op. cit., 173.
60. Ibid, 140.

CHAPTER 3

STRUCTURE OF THE SMALL INTESTINE

ALTHOUGH there may be some percutaneous absorption and some absorption from the lungs and stomach, generally speaking, minerals are absorbed into the body from the intestines. To illustrate, intestinal absorption for calcium occurs in the small intestine[1] and specifically in the upper portion of the small intestine.[2] Wilson, *et al.*, reported that magnesium is absorbed in the small intestine.[3] Most iron absorption is localized in the duodenum and jejunum where there is much higher absorption activity than in the ileum.[4] Evans[5] has reported that zinc is absorbed through the duodenum. Likewise, manganese is also bound and absorbed in the duodenum.[6] Copper, too, is absorbed from the acid pH environment of the upper portion of the gastrointestinal tract.[7]

Moog has written that nutrients, including minerals, enter the mammal's body through the mouth, but they do not become part of the body until they have crossed the lining of the small intestine and entered the bloodstream. "The interior of the gastrointestinal tract is thus more closely connected to the outer world than it is to the muscles, nerves, skeleton and blood vessels of the inner organism. The crossing of this boundary is the fundamental event in the assimilation of nutrients."[8] Thus, while a complete treatment of intestinal physiology is beyond the scope of the present work, it does become necessary to examine the small intestine briefly in order to more fully comprehend its role in mineral absorption.

The small intestine is one part of the total alimentary canal. The alimentary canal is a tube of sorts that begins at the mouth where the food is taken in and terminates with the anus with the expulsion principally of undigested food residue and sloughed cellular material. As shown by Figure 3-1, the mouth, the pharynx, the esophagus, the stomach, the small intestine, and the large intestine are all parts of the alimentary canal and play essential roles in extracting nourishment from the foods consumed.[9] Nevertheless, it is the small intestine where the majority of the actual nutrient absorption takes place.[8]

The small intestine is the longest part of the alimentary canal. It is divided into three sections: the duodenum, the jejunum, and the ileum. Entry to the small intestine is from the stomach, through the pyloric valve into the duodenum. Exit into the large intestine is from the ileum and through the ileocecal sphincter.[10]

As indicated above, the first section of the small intestine is the duodenum. In man it comprises the first 25 to 27 centimeters of intestine in length. It is longer or shorter for other mammals depending, of course, on their physical size. For example, the average length of the small intestine in an adult human being is approximately 20 feet,[10] whereas it is 130 feet in the average cow (although the small intestine in cattle functions similarly to that of nonruminants[8,11]). There are no special marks to distinguish the jejunum, but that section of the small intestine is so called because in man it is frequently empty after death. The ileum makes up about two-thirds of the small intestine.[2]

The pH of the duodenum is usually very acidic due to the low pH of the digesta, or chyme, entering the intestine from the acidic environment of the stomach. The pH of the tract increases as the digesta continues down the duodenum due to potent buffers secreted by the accessory organs of digestion. Pancreatic juice from the pancreas and bile from the liver by way of the bile duct and gall bladder are alkaline intestinal juices that are mixed with the chyme and elevate not only the pH of the digesta but also the intestinal environment as a whole.

By the time the digesta has reached the second section of the small intestine, the jejunum, the environmental pH has usually shifted to an alkaline state which increases in the ileum. This fact alone is extremely important in considering the intestinal absorption of metal ions from ingested salts.

A general requirement for absorption of a mineral derived from

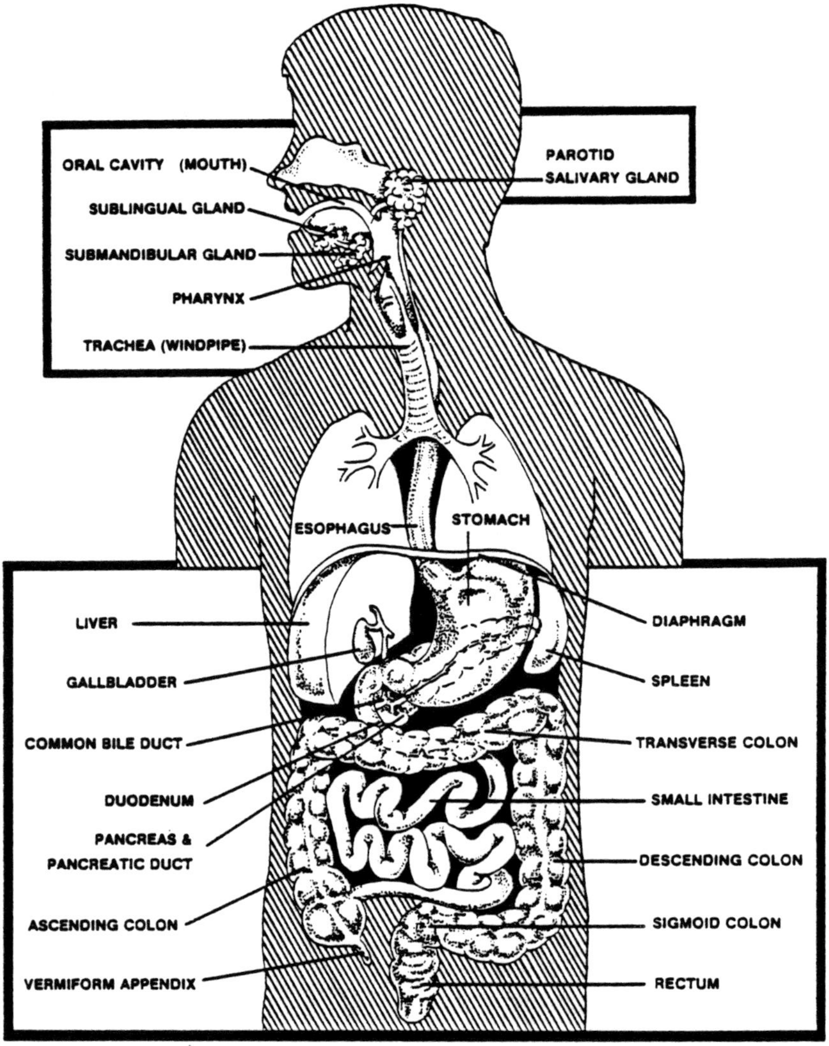

Figure 3-1. The alimentary canal of a human being.

a salt is its ability to solubilize. Minerals are not taken up by intestinal mucosal cells as intact salts but as cations, usually after being complexed. As was previously shown (Figure 2-6), there is a decreasing absorption rate as the mineral traverses the small intestine.[13] One reason for this decrease is presumably due to pH changes in the intestine. In an acid environment, most mineral salts are very soluble and easily ionized. As the minerals move down the intestine, alkaline digestive juices are introduced with a corresponding elevation in pH to alkalinity. Consequently, the solubility of these mineral ions decreases because the shift in pH tends to induce them to rejoin with free anions in the digesta.[14] Solubility results in ionization of the ingested mineral salt and/or complex. This is critical because it is only in this ionic state that it is possible to carry the mineral across the intestinal cell membrane. When the soluble state is lost, so is the mineral's ability to be absorbed.

Investigators have reported that every one of the mineral salts under consideration (those involving divalent cations) must be soluble in order to be absorbed. "The rate of absorption of metal ions can be influenced by the pH of the intestinal lumen. The pH may affect the mucosal surface of the availability of the luminal contents by altering solubility characteristics, or both effects may operate simultaneously. . . . In general, the more alkaline the lumen, the lower the rate of absorption."[14] Calcium is more soluble in an acid medium and hence is more readily absorbed from the duodenum.[15,16] Copper is rapidly absorbed when the luminal contents are still acid.[17,18] Both iron and manganese are only soluble in the acidic medium of the duodenum according to Forth and Rummel.[19] Magnesium salts, like those of calcium, are poorly absorbed in an alkaline pH.[20] The pH of the intestine also influences zinc absorption.[21] This is summarized in Figure 3-2.[67]

Structurally the small intestine must fit into a relatively small abdominal cavity while still providing a large surface area for absorption of nutrients. These requirements are met by first coiling the intestine as shown with the human intestine in Figure 3-3.[22] Three or more sets of projections (at least in larger mammals) are developed extending from the mucosa into the lumen: circular folds, villi, microvilli, and perhaps glycocalyx.

The circular folds, found only in the larger mammals are ridges from eight to ten millimeters high. They usually extend from one-half to two-thirds of the way around the inner circumference of the intestine, and as a result, they increase the intestinal surface about three times. The folds begin near the pyloric valve on the duodenum and end about the middle of the ileum.[8] They are most highly developed in the distal duodenum and the proximal jejunum[40] where a large portion of the nutrients are absorbed from the small intestine.[8]

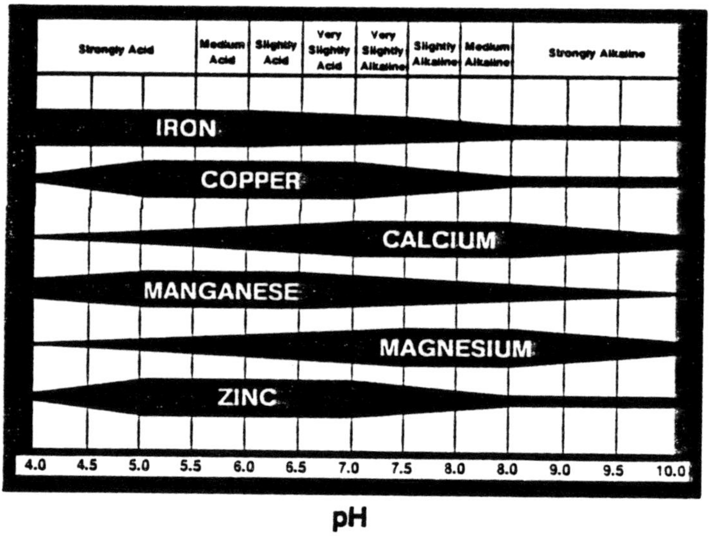

Figure 3-2. The effect of pH on solubility of minerals.

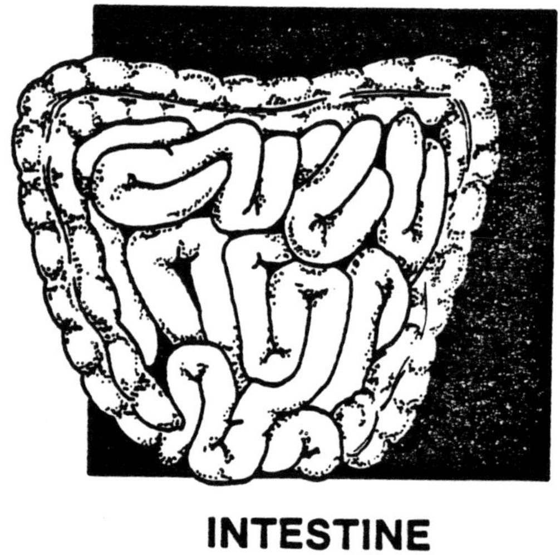

Figure 3-3. Coiling of the human intestine.[22]

The circular folds are seen in Figure 3-4.[8] Structurally the small intestine wall has the same stratified organization regardless of the segment under examination. There are three outer layers: The serosa, the muscularis (both a longitudinal and a circular muscle layer), and a submucosa. These all surround a fourth inner layer called the mucosa where actual digestion and absorption take place.[23] On the mucosa are the villi which project into the lumen from the circular folds and the surrounding mucosal surface. Each villus is covered with a single layer of absorptive epithelial cells. On the luminal side of the absorptive cell are projections called microvilli which collectively form the brush border of the absorptive cell. Finally, on the surface of the microvillus are glycocalyx, which are a network of fibers attached to the membrane of the microvilli.[8] This increases the mucosal surface of the small intestine to approximately 600 times the surface of a cylinder of similar dimensions or about 300 square meters in the average adult human being.[8,24] All of these structures are seen in Figure 3-4.[8]

As seen in Figure 3-4, the villi project into the lumen from the circular folds and the mucosa. They rise to a height of between .5 and 1.5 millimeters. Between each villus are crypts that descend into the mucosa. It is here in the "nursery" for absorptive cells, the crypts of Lieberkuhn, that new immature absorption cells are formed from undifferentiated cells dividing there. (There is some evidence that undifferentiated cells formed in the crypt also serve as precursor cells for other intestinal epithelial cells besides absorptive cells including paneth cells, endocrine cells, and most goblet cells.[50]) As these cells mature they, the absorptive cells, travel from the crypt upwards to the tip of the villus, a trip that requires from two to five days depending on the species of mammal being studied.[25] This single layer of epithelial absorption cells form an unbroken line from the crypts of Lieberkuhn (the nursery of the absorptive cells) to the tip of the villus. It is there that they are shed off into the lumen carrying with them whatever they have absorbed and not passed through to the submucosa. It is estimated in the adult human approximately 17 billion cells are sloughed off each 24 hours,[8] which amounts to about one-half pound per day. When these cells die and are sloughed off, minerals that have been absorbed into the intestinal absorptive cells and are not yet passed on to the plasma are lost to the feces[39] as seen in Figure 3-5.[41]

The cells in the crypt near the spongy mucosal floor are not as well developed as those at the tip of the villus. They are shorter, more cube shaped, and certain markings on their surface are less developed.[8] As they migrate upwards, these cells mature into true ab-

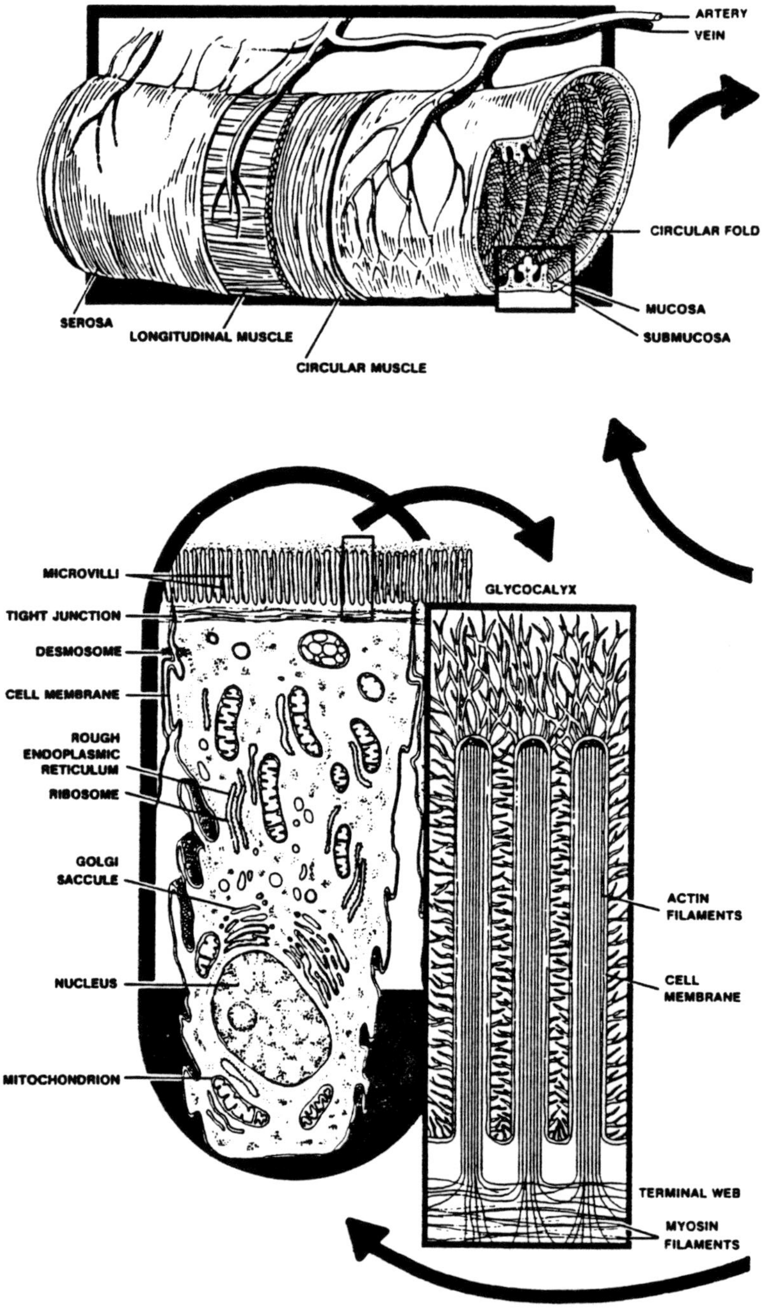

Figure 3-4A. The structures of the small intestine.[8]

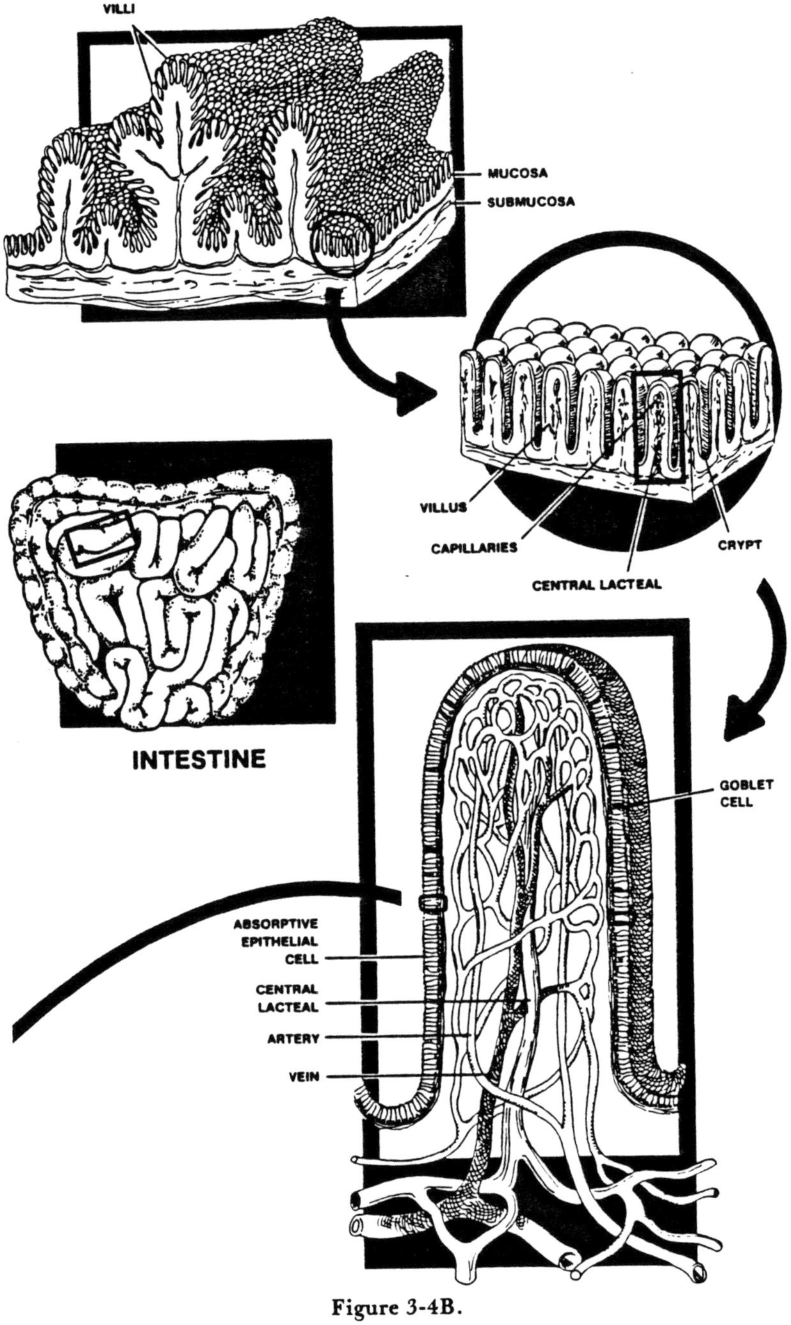
VILLI
MUCOSA
SUBMUCOSA
VILLUS
CAPILLARIES
CRYPT
CENTRAL LACTEAL
INTESTINE
GOBLET CELL
ABSORPTIVE EPITHELIAL CELL
CENTRAL LACTEAL
ARTERY
VEIN

Figure 3-4B.

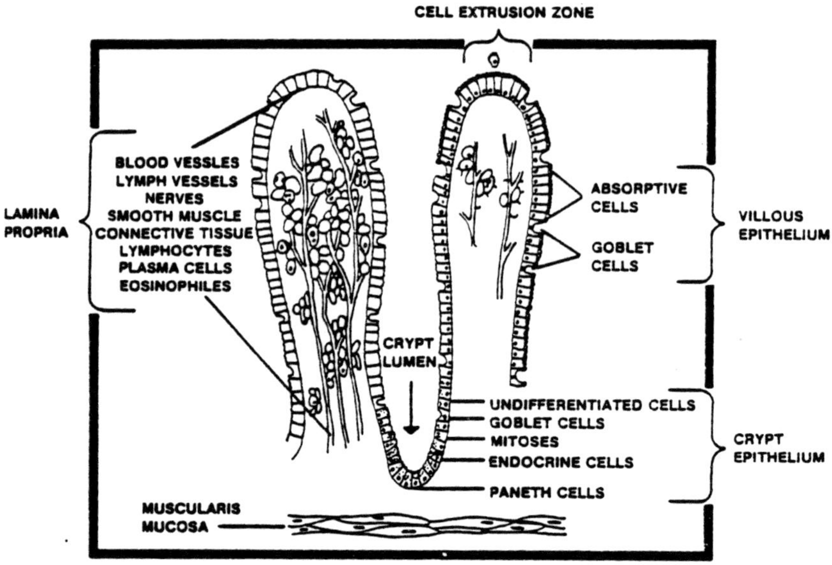

Figure 3-5. Schematic diagram of two sectioned villi and a crypt illustrating the histological organization of the mucosa of the small intestine.[41]

sorption cells becoming taller and narrower in the process[8] and at the same time developing the brush border enzymes that will be reviewed later.[25]

Besides the absorption cells there is a second type of cell embedded in the epithelial layer of the villus. This is the goblet cell. Using the process of erocytosis,[48] the goblet cells secrete a mucus that lines and protects the intestinal wall[8] through lubricating the intestine and through both its cleansing and antibacterial properties.[26,27] Chemically the mucus appears to be a glycoprotein.[49]

Before examining the absorption cells in detail, it is perhaps more logical to consider the anatomy and physiology of the villus first. As seen in Figure 3-4,[8] within the villus is a network of blood vessels and a single central lymph vessel to collect the absorbed nutrients not captured by the blood. They are both supported by a loose fibrous tissue.[8,28] The vascular anatomy of a human villus is seen in Figure 3-6.[28]

Jodal has reported that usually a single arterial vessel with a diameter of 20 micrometers transverses the center of the villus to the tip without branching. Its wall is composed of a single layer of

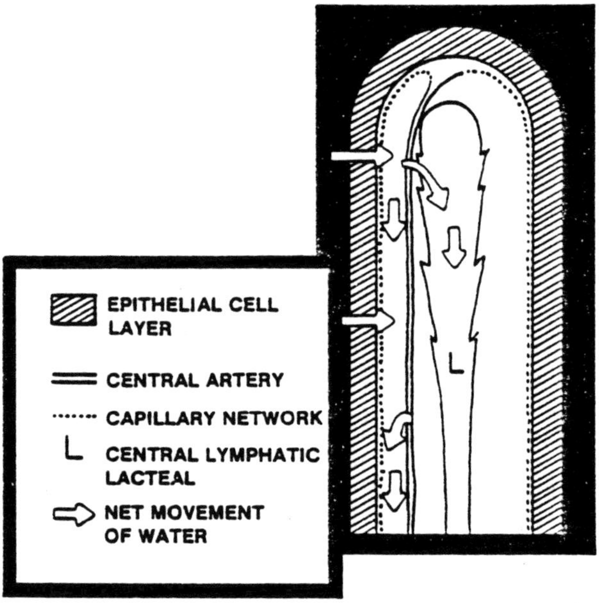

Figure 3-6. Vascular anatomy of the villus.[28]

endothelial cells. At the tip of the villus this arterial vessel branches into a dense subepithelial capillary network[29] which then course down the periphery of the lamina propria, the layer of tissue directly below the absorptive cells.[42] There are pores in these capillaries that are highly permeable and which tend to face the basement membrane of the absorptive cells so as to efficiently collect all absorbed nutrients.[42] There are at least ten times more pores in these capillaries than found in skeletal muscle capillaries.[29] At different levels of the villus these subepithelial capillaries join together to form veins. The veins then collect into descending layer veins with a flow of blood that is opposite that of the artery.[28] Thus, the artery brings oxygenated blood to the villus while the veins return the blood laden with nutrients.

As stated earlier, the luminal side of the villus is covered by a single layer of epithelial cells, mostly absorptive cells interspersed with goblet cells. The absorptive cells carry out the main functions of being both the final digestion point of nutrients as well as their absorption point from the lumen.

The absorptive cells rest upon a filamentous basal lamina or basement membrane. This in turn covers the connective tissue of the mucosa. The absorptive cells are joined together near their apical surfaces, just below their upper surface, by the zonula occludens (tight junctions) which are seen in Figure 3-4 and Figure 3-7.[31] This juncture is so tight molecules cannot penetrate between the absorptive cells from the lumen and enter the blood vessels of the villus by intercellular channels[8]. There may, however, be some passage of molecules that are expelled into the apical junction below the tight junction that move downward into the capillary and/or lymph system of the microvillus and are absorbed. It is believed by many that the tight junction is designed to prevent the reverse flow of nutrients back into the lumen rather than prevent absorption altogether.[28] There are channels on the membrane that link one cell interior to another as electrolytes of 14 angstroms or smaller are passed between cells at these junctions.[51]

The absorptive cells are fused at a number of points which, as stated above, prevents passage of nutrient molecules between these epithelial cells from the lumen. This tight junction is accomplished somewhat like the way a zipper works. There is a network of ridges on the cytoplasm of the one absorptive cell and corresponding grooves on the other cell. This effectively seals the space between the cells.[32]

Below the tight junctions is a looser connection between the absorptive cells called desmosomes. The desmosomes begin after the tight juncture at the upper end of the cells and are found intermittently to the basement membrane where they are attached to the mucosal membrane. This is also seen in Figure 3-4 and Figure 3-7.[34] At this point the absorptive cells are separated by a distance of between 10 to 30 nm. Here the plasma membrane of the cell thickens with an amorphous material which fills up the space between the cells and which also extends fibrils leading from the thickening into the cytoplasm of each absorptive cell.[33] The desmosomes which start just below the tight junction of the absorptive cells continue down as a band to just above the basement membrane. There they stop and are replaced by a half desmosome somewhat like a spot weld or rivet. Instead of functioning as a pair between the cells where fibrils from the desmosome extend equally into the adjacent absorptive cells, the half desmosome is a single unit that anchors the absorptive cell to the basement membrane.[33] The mucosal membrane also penetrates deeply into the base of the absorptive cell to facilitate rapid transport of nutrients into the blood system of the villus.[illegible] This also has the tendency to anchor the absorptive cells to the mucosal membrane.

Together with the tight junctions, the desmosomes bind the absorptive cells [illegible] that covers [illegible] villus.[8] (Figure 3-7 and Figure 3-8).

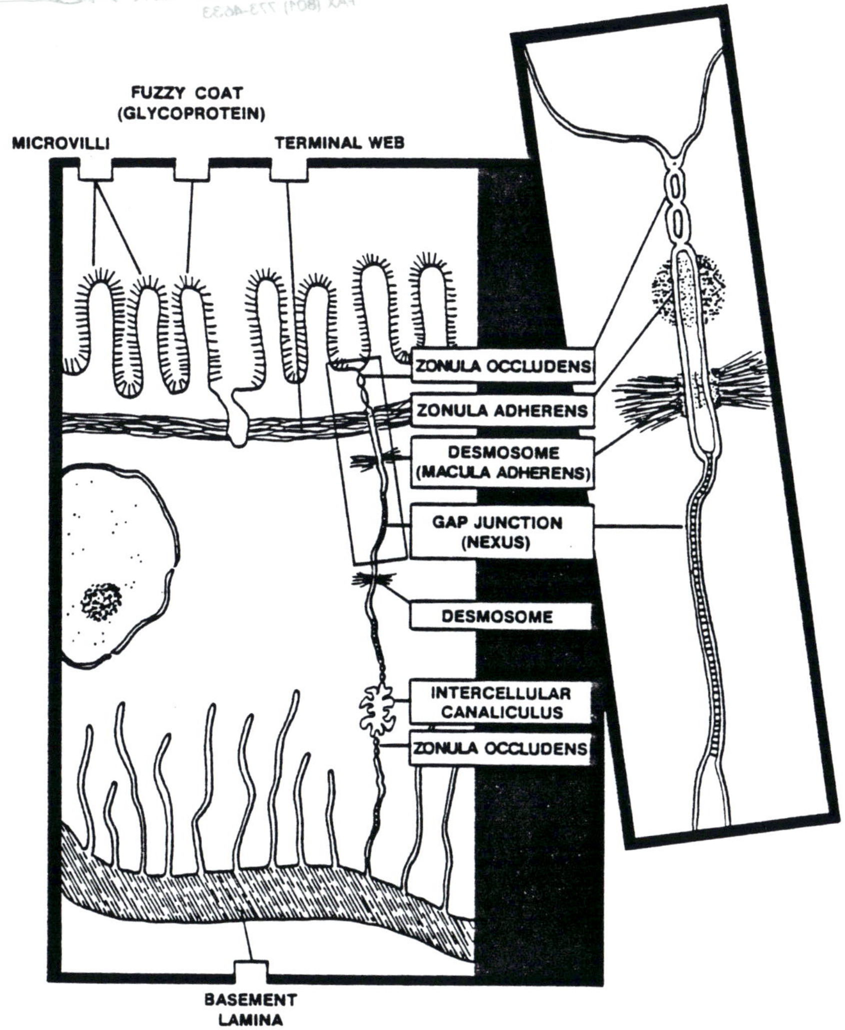

Figure 3-7. The binding of the absorptive cells.[34]

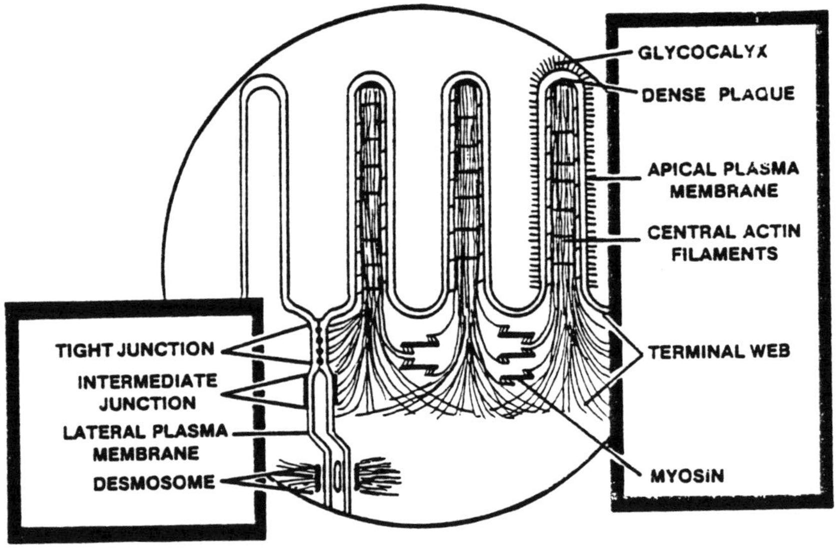

Figure 3-8. The use of the tight junction and desmosome to bind absorptive cells together.[43]

The actual interior of the absorptive cell is relatively similar to most cells although it has a higher number of energy-related organelles presumably for the active transport of nutrients from the lumen into the absorptive cells. As seen in Figure 3-4, the nucleus is located deep in the cytoplasm towards the basal third of the cell. Above it is a zone that is richly endowed with mitochondria, which supply energy for the specialized operation of the cell, endoplasmic reticulum, where proteins are assembled, and Golgi saccules, where carbohydrate side chains are attached to proteins, which are necessary for the very high oxidative and synthetic activity demanded of the absorptive cell. Furthermore, each of these organelles is able to concentrate specific metals before moving them to the basement membrane and the blood.[36] It is absolutely necessary that a large supply of energy be generated at all times in order to facilitate the active transport of nutrients from the lumen into the absorptive cells. A detailed description of the transport of nutrients into each of the organelles is outside the scope of this work.

The luminal end of the absorptive cell, called the brush border, is composed of projections that are 0.6 to 0.8 micrometers long and 0.1 micrometer in diameter.[35] They are the microvilli and increase the

Each one of the protein rods seen in Figure 3-11, is an alpha helix. The backbone of the chain is coiled tightly into a helix which forms a cylinder. The amino acid side chains project from the cylinder. Each helix is stabilized through hydrogen bonding between the hydrophilic NH and CO groups of the backbone chain. This bonding occurs between the successive turns of the alpha helix. Where the helixes are linked together the stabilizing bonding may not occur.[77]

As mentioned above, this integral protein is extremely important because it is the carrier or channel for the active transport of hydrophilic molecules, such as cations, across the cell membrane.[56] The exact process for transport will be dealt with in greater detail in Chapter 6. The membrane proteins give mechanical structure to the membrane and are involved in regulatory or ligand recognition properties of the membrane.[61] Rather than examining each function in depth, at present it is only important to realize that the membrane of the absorptive cell has a very high concentration of integral proteins embedded within it when compared to the membrane of other types of cells. This is clearly seen in Table 3-1.[57,58]

This high protein-to-lipid ratio in the absorptive cell membrane correspond to the large number of intramembrane particles seen with an electron microscope when preparations of small intestine epithelial cells have been frozen and then fractured. This technique results in membrane cleavage along its internal hydrophobic core allowing visualization of its structural features. The intramembrane particles are distributed in an asymmetric fashion between the inner and outer membrane faces.[59] There are far more of these particles on the luminal side of the absorptive cell membrane than on the basolateral membrane or on other plasma membrane surfaces,[60] all of which further suggest the very important role of the integral proteins as transporters in the absorption of nutrients from the lumen into the absorptive cells lining the villus.

The arrangement of the proteins in the membrane of the cell helps to explain why the different enzymes involved in intestinal absorption may have their active sites exposed to the outer surface of the membrane such as seen with glycolipid and glycoprotein composed glycocalyx on the microvilli. Through functioning of these enzymes[62] and the dynamic fluid structure of the membrane, proteins of appropriate size or groups of protein subunits may pass across the membrane.[63,64] This would allow the traversing protein to remain in

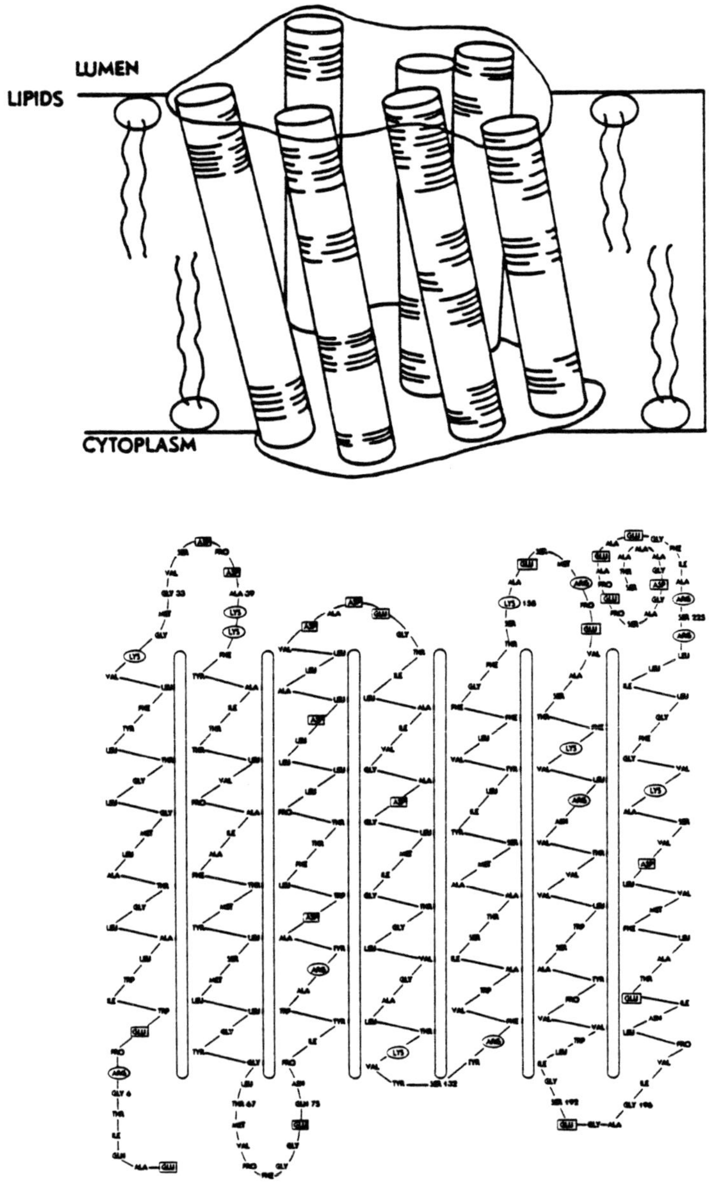

Figure 3-11. A schematic view of the disposition of the seven alpha-helices of bacteriorhodopsin in the lipid bilayer of the membrane. At the right is the suggested arrangement of the polypeptide as it transverses the membrane. Positively charged amino acids are encircled and the negatively charged amino acids are in rectangular boxes.

Figure 3-12. The proposed three-dimension structure of gap junctions in the membrane. The model shows that a change in the angle of tilt would open and close the pore and allow the absorbed molecule to be moved along it by being passed from amino acid that forms the helices (integral protein).

Table 3-1

LIPID AND PROTEIN RATIOS IN SOME CELL MEMBRANES

Species and Tissue	*Protein*	:	*Lipid*
Human Absorptive Cell	1	:	0.59
Human CNS Myelin	1	:	3.95
Human Erythrocyte	1	:	0.67

contact with the aqueous solvent on both sides of the membrane thus fulfilling the hydrophilic requirement for active transport by integral protein carriers.[56] This is seen in Figure 3-13[65] in greater detail than the previous figure.

Van Deenen[73] has indicated that the permeability of the membrane to cations is dependent upon the fatty acid content in the membrane. If it contains long saturated hydrocarbon chains, then permeability is reduced. Unsaturated hydrocarbon chains tend to loosen the membrane structure which increases its permeability. An increase in the steroid content of the membrane results in decreased permeability. "The phosphate group of phospholipids may be bound

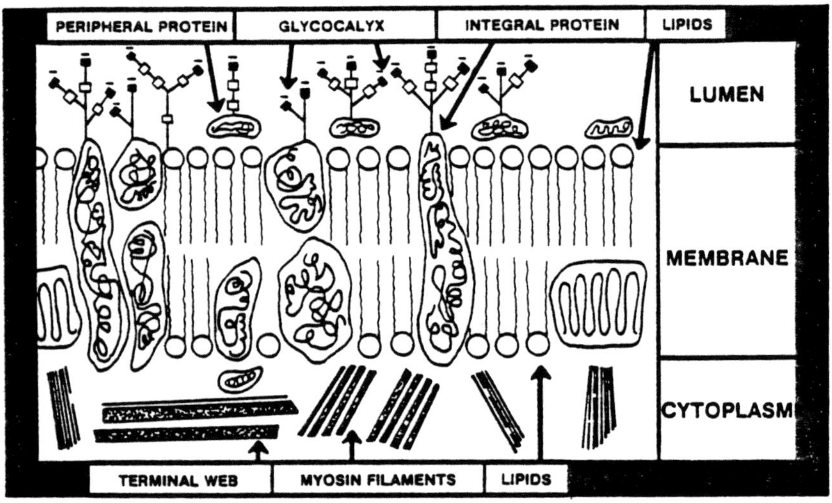

Figure 3-13.[57] Cross section of a mucosal cell membrane showing glycocalyx.

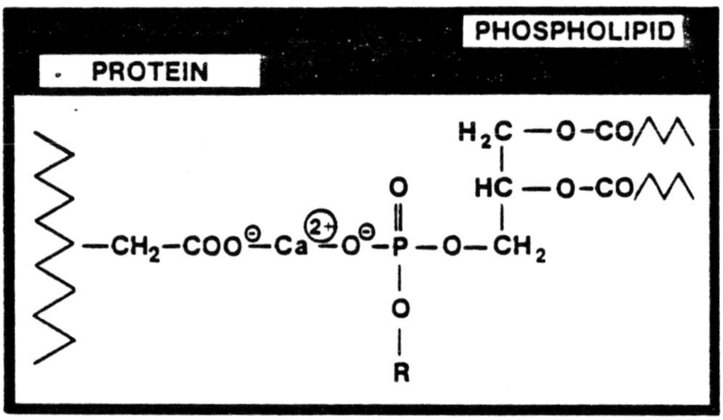

Figure 3-14. The permeability of the membrane being influenced by Ca^{+2} as described in the text.

to the NH_3^+ group of proteins by electrostatic forces. In addition the phosphate group can be bridged by Ca^{2+} to a carboxylic group of proteins [Figure 3-14] . . . In an acid medium Ca^{2+} is replaced by H^+ and this bond is broken thus drastically increasing the permeability of the membrane."[70]

In summary, it can be seen from this brief review the extremely important role played by the small intestine and the complex design of the mucosa which is necessary in order to carry out the absorptive functions. "Material absorbed from the lumen must first traverse the epithelium to gain access to mucosal blood and lymph vessels for distribution to more distant tissues, just as material delivered by the intestine to the lumen must either traverse or be secreted by the epithelium. Thus, this epithelium must regulate diverse absorptive secretory processes. But more than that, during absorption many nutrients are modified by synthetic and degradative processes as they cross the intestinal epithelium, and many of the secretions delivered into the intestinal lumen are actually synthesized and assembled within intestinal epithelial cells. It is therefore not surprising that the epithelium of the small intestine is a highly differentiated structure whose constituent cells are endowed with complex morphological specializations that facilitate their diverse functions."[66]

Thus it can be seen that the barrier which controls the flux between the intestinal lumen and lamina propria is more than an inert membrane. Instead it is a complex dynamic organ which is composed of living cells.[31]

REFERENCES

1. Schachter, D., *et al.*, "Active Transport of Calcium by the Small Intestine of the Rat," Amer. J. Physiol., 198:263, 1960.
2. Wilson, E., *et al., Principles of Nutrition* (New York: John Wiley & Sons), 262, 1979.
3. Ibid., 273.
4. Forth, W. and Rummel, W., "Absorption of Iron and Chemically Related Metals *in vitro* and *in vivo*: Specificity of the Iron-Binding System in the Mucosa of the Jejunum," in Skoryna, S. and Waldron-Edward, E., eds., *Intestinal Absorption of Metal Ions, Trace Elements and Radionuclides* (Oxford: Pergamon Press), 174, 1971.
5. Evans, G., "Zinc Absorption and Transport," in Prasad, A., ed., *Trace Elements in Human Health and Disease* (New York: Academic Press), V1, 183-185, 1976.
6. Thompson, A., *et al.*, "Interrelation of Intestinal Transport System for Manganese and Iron" J. Lab. Clin. Med., 78:642, 1971.
7. Thompsett, S., "Factors Influencing the Absorption of Iron and Copper from the Alimentary Tract," Biochem. J., 34:961, 1940.
8. Moog, F., "The Lining of the Small Intestine," Scientific American, 245:154, November, 1981.
9. Memmler, R., and Wood, D., *Structure and Function of the Human Body* (Philadelphia: J.B. Lippincott Co.), 139, 1977.
10. Ibid, 143.

11. Miller, W., *Dairy Cattle Feeding and Nutrition* (New York: Academic Press), 14, 1979.
12. Davidson, A., *Mammalian Anatomy* (Philadelphia: P. Blakisten's Sons & Company, Inc.), 158, 1937.
13. Waldron-Edward, D., *et al.*, "Effects of the Counter Ion and pH on Intestinal Absorption of Calcium and Strontium," Proc. Soc. Exp. Biol. Med., 123:532, 1966.
14. Waldron-Edward, D., "Effects of pH and Counter-Ion on Absorption of Metal Ions," in Skoryna, S. and Waldron-Edward, D., eds., *Intestinal Absorption of Metal Ions, Trace Elements and Radionuclides* (Oxford: Pergamon Press), 381, 1971.
15. Wilson, op. cit., 262.
16. Guthrie, H., *Introductory Nutrition* (St. Louis: C.V. Mosby Co.), 116, 1975.
17. Ibid., 174.
18. Cartwright, G., *et al.*, "Studies on Copper Metabolism," J. Clin. Invest., 33:1487, 1954.
19. Forth, op. cit., 181.
20. Briggs, G. and Calloway, D., *Nutrition and Physical Fitness* (Philadelphia: W.B. Saunders), 265, 1979.
21. Oelshlegel, F., and Brewer, G., "Absorption of Pharmacological Doses of Zinc," in Brewer, G. and Prasad, A., eds., *Zinc Metabolism* (New York: Alan R. Liss), 310, 1977.
22. Guyton, A., *Textbook of Medical Physiology* (Philadelphia: W.B. Saunders Co.), 770, 1971.
23. Specht, W., "Morphology of the Intestinal Wall," in Kramer, M. and Lauterbach, F., eds., *Intestinal Permeation* (Amsterdam: Excerpta Medica), 4, 1977.
24. Vaughan, B., "Intestinal Electrolyte Absorption by Parallel Determination of Unidirection Sodium and Water Transfers," Am. J. Physiol., 198:1235, 1960.
25. Hendrix, T., "The Absorptive Function of the Alimentary Canal," in Mountcastle, V., ed., *Medical Physiology* (St. Louis: C.V. Mosby Co.), V2, 1145-7, 1974.
26. Moe, H., "On Goblet Cells, Especially of the Intestine of Some Mammalian Species," Int. Rev. Cytol., 4:229, 1955.
27. Subbuswamy, S., "Patterns of Mucin Secretion in Human Intestinal Mucosa", J. Anat., 108:291, 1971.
28. Jodal, M., "The Intestinal Countercurrent Exchanger and its Influence on Intestinal Absorption," in Kramer, M. and Lauterbach, F., eds., *Intestinal Permeation* (Amsterdam: Excerpta Medica), 48, 1977.
29. Biber, B., *et al.*, "Intramural Blood Flow and Blood Volume in Small Intestine of the Cat as Analyzed by an Indicator-Dilution Technique," Acta. Physiol. Scand., 87:391, 1973.
30. Barrowman, I. and Roberts, K., "The Role of the Lymphatic System in the Absorption of Water from the Intestine of the Rat," Quart. J. Exp. Physiol., 52:19, 1967.
31. Rubin, W., "The Epithelial Membrane of the Small Intestine," J. Clin. Nutr.,

24:45, 1971.
32. Staehelin, L. and Hull, B., "Junctions Between Cells," Scientific Amer., 238:141, May, 1978.
33. Kelly, D., "Fine Structure of Desmosomes, Hemedesmosomes and an Adepidermal Globular Layer in Developing New Epidermis," J. Cell. Biol., 28:51, 1966.
34. Robertis, E., *et al.*, *Cell Biology* (Philadelphia: W.B. Saunders Co.), 159, 1975.
35. Ibid., 158.
36. Giese, A., *Cell Physiology* (Philadelphia: Saunders College), 200, 1973.
37. Kinter, W. and Wilson, T., "Autoradiographic Study of Sugar and Amino Absorption by Everted Sacs of Hamster Intestine," J. Cell. Biol., 25:19, 1965.
38. Robertson, J., Brochem. Soc. Symp., 16:3, 1959.
39. Briggs, op. cit., 279.
40. Trier, J. and Madara, J., "Functional Morphology of the Mucosa of the Small Intestine," in Johnson, L., *et al.*, ed., *Physiology of the Gastrointestinal Tract* (New York: Raven Press), V2, 925, 1981.
41. Ibid, 926.
42. Ibid., 927.
43. Ibid., 930.
44. Mukherjee, T. and Stainhelin, L., "The Fine-Structural Organization of the Brush Border of Intestinal Epithelial Cells," J. Cell. Sci., 8:573, 1971.
45. Brestcher, A. and Weber, K., "Localization of Actin and Microfilament-Associated Proteins in the Microvilli and Terminal Web of the Intestinal Brush Border by Microscopy," J. Cell. Biol., 79:839, 1978.
46. Trier and Madara, op. cit., 939.
48. Specian, R. and Neutra, M., "Acceleration of Secretion in Colonic Goblet Cells by Acetylcholine," J. Cell. Biol., 85:626, 1980.
49. Forstner, G., *et al.*, "Goblet Cell Nucin of Rat Small Intestine. Chemical and Physical Characterization," Can. J. Biochem., 51:1154, 1973.
50. Cheng, H. and Lebland, C., "Origin, Differentiation and Renewal of the Four Main Epithelial Cell Types in the Mouse Small Intestine. II. Unitarian Theory of the Origin of the Four Epithelial Cell Types," Am. J Anat., 141:537, 1974.
51. Antonov, V., *et al.*, "The Appearance of Single-Ion Channels in Unmodified Lipid Bilayer Membranes at the Phase Transition Temperature," Nature, 283:585, 1980.
52. Singer, S. and Nicolson, G., "The Fluid Mosaic Model of the Structure of the Cell Membrane," Science, 175:720, 1972.
53. Gitler, C. and Montal, M., "Thin-Proteolipid Films: A New Approach to the Reconstitution of Biological Membranes," Biochem. and Byophy. Res. Comm., 47:1486, 1972.
54. Sollner, K., "Ion Transport Across Membranes of High Ionic Selectivity and Transmissivity and Their Use in Model Studies of Biophysical Significance," in Skoryna, S. and Waldron-Edward, D., eds., *Intestinal Absorption of Metal*

Ions, Trace Elements and Radionuclides (Oxford: Pergamon Press), 40, 1971.
55. Giese, op. cit., 192.
56. Wilson, D., "Cellular Transport Mechanisms," Ann. Rev. Biochem., 47:933, 1978.
57. Triggle, D., *Neurotransmitter-Receptor Interactions* (New York: Academic Press), 122, 1971.
58. Trier and Madara, op. cit., 931.
59. Branton, D., "Fracture Faces of Frozen Membranes," Proc. Natl. Acad. Sci., 55:1048, 1966.
60. Madara, J., *et al.*, "Structural Changes in the Plasma Membrane Accompanying Differentiation of Epithelial Cells in Human and Monkey Small Intestine," Gastroenterology, 78:963, 1980.
61. Robertis, op. cit., 147.
62. White, A., *et al.*, *Principles of Biochemistry* (New York: McGraw Hill Book Co.), 633, 1973.
63. Steck, T., *et al.*, "Disposition of the Major Proteins in the Isolated Erythrocyte Membrane Proteolytic Dissection," Biochem., 10:2617, 1971.
64. Bretscher, M., "Human Erythrocyte Membranes: Specific Labelling of Surface Proteins," J. Mol. Biol., 58:775, 1971.
65. Pollock, E., "Fine Structural Analysis of Animal Cell Surfaces: Membrane and Cell Surface Topography," Am. Zool., 18:25, 1978.
66. Trier and Madara, op. cit., 925.
67. Hsu, H. and Graff, D., "Uptake of Micronutrients by Plants," in Ashmead, D., ed., *A New Era in Plant Nutrition* (Clearfield: Albion Laboratories, Inc.), 41, 1982.
68. Singer, S., "A Fluid Lipid-Globular Protein Mosaic Model of Membrane Structure," Ann. N.Y. Acad. Sci., 195:16, 1972.
69. Seelig, J., "Structure of Lipids in Biological Membranes," paper presented at Biophys. Colloq., Gissen, 1980.
70. Mengel, K. and Kirkby, E., *Principles of Plant Nutrition* (Bern: International Potash Institute), 111-114, 1980.
71. Lodish, H. and Rothman, J., "The Assembly of Cell Membranes," Sci. American, 240:38, 1979.
72. Walker, N., "The Structure of Biological Membranes," in Luttge, U. and Pitman, M., eds., *Encyclopedia of Plant Physiology* (Berlin: Springer-Verlag), V2, 3-11, 1976.
73. Van Deenen, L., "Phospholipids. Relationships Between Their Chemical Structure and Biological Membranes," Naturwissenschaften, 59:485, 1972.
74. Warren, G., "Membrane Proteins: Structure and Assembly," in Finean, J. and Michell, R., eds., *Membrane Structure* (Amsterdam: Elsevier), 251, 1981.
75. Ibid,., 226.
76. Flagg-Newton, J. and Loewenstein, W., "Asymmetrically Permeable Membrane Channels in Cell Junction," Science 207:771, 1980.
77. Unwin, N. and Henderson, R., "The Structure of Proteins in Biological Membrane," Scientific American, 250:78, Feb 1984.

molecules or ionic molecules; and (5) the coordinate covalent bonds have definite spatial arrangements.[38]

In discussing metal-ligand bonds, Craig and Nyholm[1] summarized them as follows:

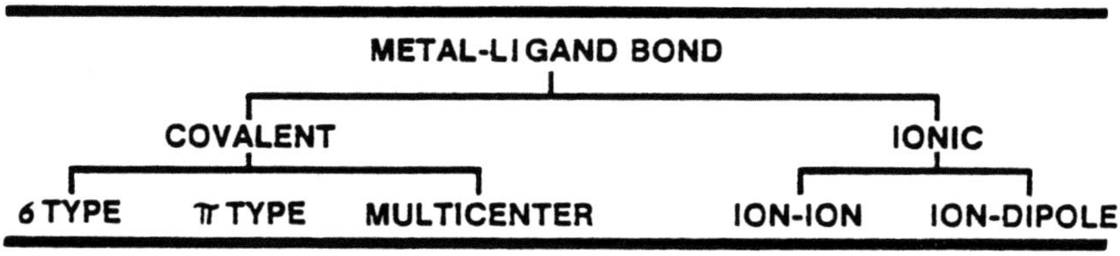

Figure 4-3. Metal ligand bonds.

In the use of ionic bonds the metal can be attached to the ligand by means of electrostatic attraction between oppositely-charged ions. The metal ion is thusly attracted to the ligand due to its excess of electrons seen as its negative charge. There is a "sharing" of electrons by both the metal and the ligand in a covalent bond. If the ligand has a neutral charge, such as a NH_3 molecule, an ion-dipole bond is formed. It is both a static and an induced attraction between the metal ion and the ligand. Such a bond is seen in hexammine-cobalt (III) ions in Figure 4-4.

Another examle can be seen in an aqueous solution containing a metal ion. The water molecules are strongly bound to the metal with the negative oxygen of the H_2O molecule dipole oriented towards the

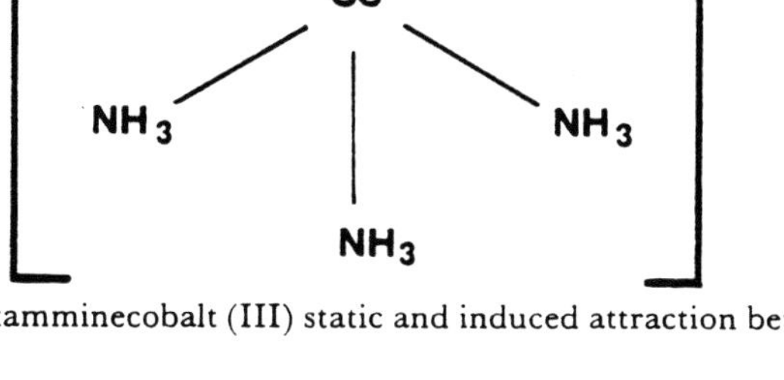

Figure 4-4. Hexamminecobalt (III) static and induced attraction between the ion and ligand.

positive-charged metal ion.[2] The water molecules can be replaced or displaced by other groups, including chelating agents. In all cases, the groups bonded to the ion can be considered as ligands.

In covalent bonding there is a donation of electrons to the ion by the ligand. The metal ion acts as a Lewis acid and as such is an electron pair acceptor. The ligand functions like a Lewis base and is the electron pair donor.[3] As such they have unshared electron pairs or electron pairs that can be made available to be bound to or shared with the cation.[4] If both electrons are donated by the ligand, the bond is considered a coordinate covalent bond. If one electron is furnished by the metal and the other by the ligand, then the resulting bond is a normal covalent bond.[5] A coordinate covalent bond is required for chelation.

"In the extreme ionic bond we suppose that a metal ion M^{n+} is attached to a certain number of ligands, either charged, like Cl^-, or uncharged, like NH_3. If the ligands are negatively charged, the model is held together by electrostatic attraction between oppositely charged ions. If the ligands are neutral, the attraction is between the M^{n+} ion and the dipole (static plus induced) on the NH_3 groups. In the extreme covalent bond, the binding is essentially of the type familiar in the covalent bonds between hydrogen atoms in H_2 or carbon atoms in hydrocarbons; however, such a picture cannot apply at all exactly, because the bonded atoms in a complex are very different in size, and the better analogy would be the heteronuclear C-H bond. Also, the bonded atoms are of different electronegativities, so that in most cases the bond will have some polar character. With these points in mind, we subdivide the covalent bond into three types: first, the σ bond in which the electron density is greatest along the internuclear axis, and formed in many cases, for example by the overlapping of a lone pair orbital of the ligand with a dsp hybrid orbital of the metal; second the π bond analogous to the auxiliary bond in ethylene, formed by the overlapping of π-type orbitals on ligand and metal as in compounds of the type

$$(CH_3)_2\overset{-}{B} = \overset{+}{N}(CH_3)_2$$

and in the carboxyls of nickel and other transition metals; and, finally, other bonding situations that may broadly be described as multicenter bonds. The most familiar examples are in electron-deficient molecules such as B_2H_6 with a single pair of electrons concentrated in the region of overlap of an sp^3 hybrid orbital from each boron atom and the orbital of the hydrogen atom. Among metal complexes the metal cyclopentadienyls are examples of multicenter bonding, insofar as the bond forces cannot be treated as localized between particular atom pairs."[6]

For a ligand to be able to chelate a metal ion it must contain at least two donor atoms capable of combining with the metal ion by donating a pair of electrons.[7] The atoms within the ligand that act as donors are the more electronegative ones on the right-hand side of the periodic table. Both Group V (nitrogen, phosphorus, etc.) and Group VI (oxygen, sulfur, etc.) elements are the primary atoms that function as donors. Nitrogen, oxygen and sulfur are the donor atoms in amino acids. The halogens do not act as donor atoms in a ligand.[8]

The donating atoms within the ligand must be located within the molecule so that they permit the formation of a ring with the metal ion.[7,8] "This type of ring has been given a special name—chelate ring—and the molecule or ion from which it is formed is known as a chelating agent. The process of forming a chelate ring is known as chelation—any metal complex in which one or more chelate rings are present is defined as a metal chelate."[9]

Even though there are sufficient donor atoms within the ligand molecule, unless at least two of them are sterically capable of bonding and forming a ring, by definition, no chelation is possible. To illustrate, one donor group may easily attach itself to the cation, but the bonding by the second or third group may be prevented by a clash between the first group and the second or third group when the latter bends and moves into the correct position to attach to the cation.[7]

This clash may be brought about by the size of the metal ion. The smaller the cation, the greater the likelihood of problems, or at least the lower the stability of the chelate.[7] Furthermore, there is a natural modification of the bond angles and bond lengths of the ligand during the chelation process.[7] If the donor groups are so bulky that close approach to the cation is hindered or prevented, then it interferes with the chelation process.[8]

Chelates can vary in the size of the ring depending on the number of elements composing the ring. Three-member chelate rings are not stable.[18] The smallest chelate ring that has been established with certainty is a four-member chelate ring. One such example is the cisdihydroxytetrammine cobalt (III) ion ($[Co(NH_3)_4(OH_2)]^+$) first discussed by Werner[10] and seen in Figure 4-5.

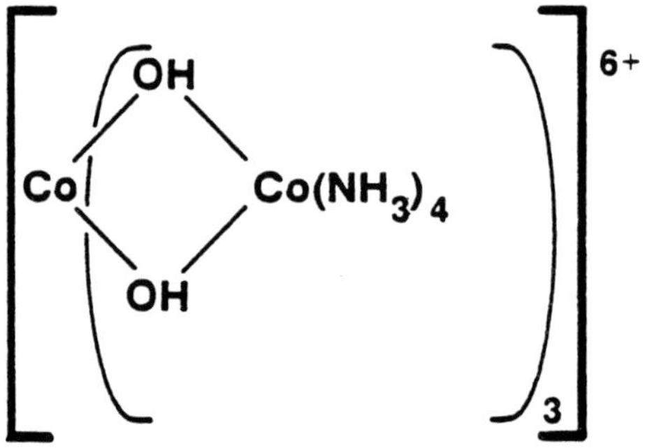

Figure 4-5. Werner's cisdihydroxytetrammine cobalt (III) ion.[10]

Generally speaking, a chelate comprised of a four-member ring is not extremely stable due to the considerable strain on the angles of the bonds forming the chelate. Using X-Ray crystal analysis it was shown that the angle of the metal bonds is much less than in a larger chelate ring and the length of the bonds is also shorter.[13] Consequently, its formation is relatively rare.[11]

Chelates comprised of five- and six-member rings are the most common chelates. The binding of Zn^{+2} to glycine at the NH_2 and O sites produces a five-membered ring. The binding of that same zinc ion to cysteine at the S and O sites results in a six-member ring although the utilization of the NH_2 donor group in the cysteine ligand will produce a five-member ring. As a general rule, five-member rings are the most stable of the chelate rings.[11] Possible combinations of five and six-member chelate rings using one amino acid of cysteine as the ligand and zinc as the metal are seen in Figure 4-6.[12]

Larger chelate rings have been studied, although not extensively. "Seven-, eight-, and higher-membered rings are doubtless formed by the higher homologs of some sexadentate molecules. . ."[11]

The two donor groups within the ligand molecule can both be acidic (COOH, SH, SO_3H, etc.) or basic (O, NH_2, NH, N, OH, S, etc.) or one donor can be acidic while the other is basic.[11] Where both donor groups are acidic, they include inorganic acids and many

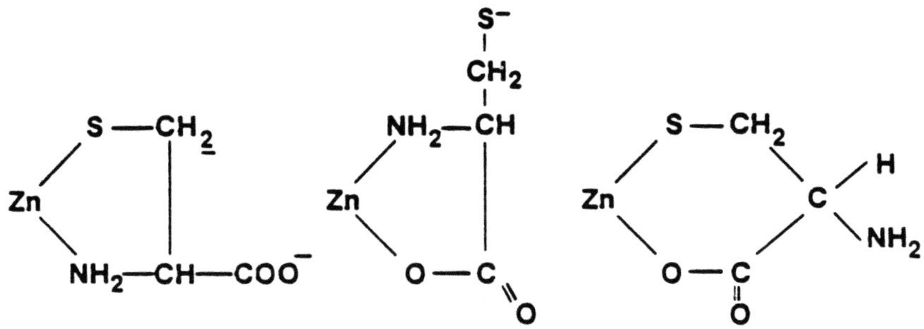

Figure 4-6. Cysteine amino acid chelates of zinc.

organic ligands. The stability of the chelate with the metal is found to decrease progressively as the number of methylene groups in the acid increases.[14]

When considering ligands in which both donor groups are basic, the donor groups are most frequently nitrogen atoms. When both donor groups are basic, they coordinate as a neutral molecule and the resulting chelate has the same charge as the original cation.[22] These types of chelates are among the most stable formed, particularly when a five-member ring is formed. Chelate rings which are formed from basic donor groups that contain six or more atoms in the rings are not as stable as the five-member ring.[15]

The most common combination of donor groups involve one acidic and one basic group. This results in a great number of ligands. Although they do not necessarily always have an acidic and basic group, one common group of ligands is the amino acids. "The simplest ligand of this kind is glycine, NH_2CH_2COOH. In aqueous solution between pH=4 and pH+9, this exists as the zwitterion, $H_3{}^+NCH_2CO_2{}^-$; this loses a proton from the $-NH_3{}^-$ group and chelates a metal ion to form a five-membered ring. This acid and other amino acids are components of protein molecules which consist essentially of α-amino acids linked together by peptide (-CONH-) linkages. Terminal $-NH_2$ and $-CO_2H$ groups are present, together with peptide linkages and side-chains. These are all potential sites for the coordination of metals. Two amino acid components of proteins with a particular affinity for metal ions are histidine and cysteine. As well as $-NH_2$ and $-CO_2H$ groups histidine has an imidazole ring and cysteine a thiol group, both of which can participate in bonding to a metal. In some structures, known as metaloproteins, the metal ion and protein are very firmly linked together so that the metal is an integral part of the molecular structure. Complexes of this type play vital roles in many biological processes."[16]

"The donor groups of ligands most commonly involved are oxygen, nitrogen, and/or sulfur. These are relatively small and highly electronegative atoms, i.e., they have the capacity to closely hold electrons. The donor groups can appear in organic substances such as carbonate, sulfite, water, or ammonia. Most commonly they are found in organic structures. Ligands may be acidic (anionic) or neutral (coordinating) groups. Both kinds or either may be involved simultaneously with a given metal. Neutral groups involved in coordination besides water are hydroxy groups, carbonyls, and -SH groups."[17]

Generally the factors that govern the bonding of an acid group to a cation are similar to those governing the bonding of it to a proton.[20] "At low pH, H^+ competes with metal ions for donor groups and at high pH, hydroxyl ions compete with donor groups for metal ions."[21] The metal ions behave as Lewis acids, that is, electron acceptors. On the other hand, the ligand behaves as a Lewis base.[26,29] This means that the more basic the donor group is, the greater the stability of the chelate.[20] For example, in solutions that have a very low pH a potential chelating ligand may attach itself to the metal ion by only one ligand atom as seen in Figure 4-7[7] and not form a chelate.

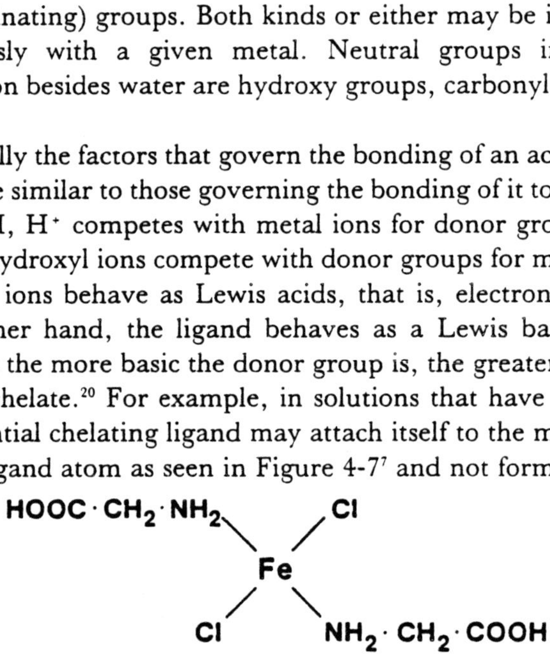

Figure 4-7. A metal complex.

The solution stability of the chelate is defined by its stability constant. The latter is an equilibrium constant for the reaction of a hydrated metal ion with the coordinating molecule to form the chelate.[23] It is a ratio of the velocity of the formation of a chelate on the one hand compared to the speed at which it is dissociating into cations and ligands on the other.[24] The higher the stability constant, a logarithmic expression, the more stable the chelate is.

Besides the basic strength of the chelating molecule, there are several factors which will affect the stability of the chelate. As mentioned earlier, the size of the chelate ring influences its stability. Three-membered ring chelates are not stable.[18] Four-membered rings are certainly more stable.[11] Five-member rings appear to be the most stable of all the chelates.[25] Past the five members, the stability of the chelate tends to decrease again. "As the number of atoms in the ring increases, the relative stability of a metal chelate compared with an analogous complex in which the ring does not exist decreases markedly."[25]

Another factor influencing stability is the number of chelate rings formed with a single metal ion.[25] The greater the number of rings, the greater the stability. In nature the more stable chelates form more than one ring, such as the formation of four rings to one iron ion in the heme molecule. Multidentate chelates, chelates in which there are more than two donor atoms in the ligand, are more stable because the ligand is able to occupy a larger number of positions in the coordination shell of the cation.[14] It should be remembered, however, that not all chelates in nature are designed to be stable. As will be seen, the integral proteins that transport cations across the mucosal membrane are bidentate chelates and form a single ring with the metal ion. On the other hand, if the metal is absorbed from the gut as a dipeptide-like amino acid chelate, it is a more complex bidentate chelate. It forms two bidentate rings with two amino acids.[26] If it were any larger, it could not be absorbed intact.

Amino acids as chelating ligands were first studied in 1904 by Ley. Using glycine he chelated copper. The resulting molecule had a molar ratio of two ligands to one atom of metal. He found through conductivity measurements that the copper ceased to be ionized and had become part of a double ring structure involving both glycine molecules such as seen in Figure 4-2.[39]

In chelating ligands the α-amino acids are of great biological importance. Over twenty of these amino acids are found in nature. This is seen in Table 4-1 along with two imino acids.[28] Each amino acid has at least two ionizable groups which are the amino (NH_2) and carboxyl (COOH) groups which is also shown in this table.

The amino acids are linked together by peptide bonds (-CONH-). Each peptide has a terminal $-NH_2$ and -COOH group because the α-carboxyl and amino functions are involved in the

Table 4-1.

AMINO ACID FOUND IN NATURE THAT CAN CHELATE CATIONS

NATURALLY OCCURRING AMINO ACIDS $R-CH(NH_2)-COOH$

R		R	
$H-$	GLYCINE		
CH_3-	ALANINE	$NH_2(O=)CCH_2-$	ASPARAGINE
$(CH_3)_2CH-$	VALINE		
$(CH_3)_2CHCH_2-$	LEUCINE	$\overset{+}{N}H_3(CH_2)_3CH_2-$	LYSINE
$CH_3CH_2CH(CH_3)-$	ISOLEUCINE	$(H_2N)_2C{=}NH(CH_2)_2CH_2-$	ARGININE
$HOCH_2-$	SERINE		
$CH_3CH(OH)-$	THREONINE	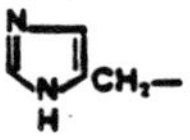	HISTIDINE
	TYROSINE	$\overset{+}{N}H_3CH(CH_2)_2CH_2-$ with OH	HYDROXYLYSINE
		$SHCH_2-$	CYSTEINE
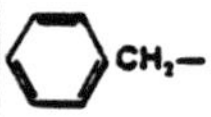	PHENYLALANINE	$CH_3SCH_2CH_2-$	METHIONINE
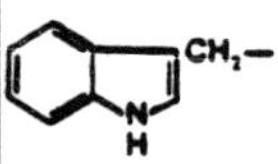	TRYPTOPHAN	$\overset{+}{N}H_3(^-OOC)CHCH_2SSCH_2-$	CYSTINE
$^-OOCCH_2-$	ASPARTIC ACID		PROLINE (IMINO ACID)
$^-OOCCH_2CH_2-$	GLUTAMIC ACID		
$NH_2(O=)CCH_2CH_2-$	GLUTAMINE	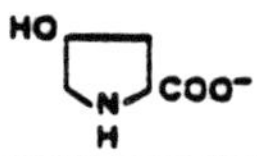	HYDROXYPROLINE (IMINO ACID)

formation of the peptide link.[28] This linkage, in the case of peptides as well as terminal $-NH_2$ and COOH groups and side chains, are all potential sites for coordination of metals.[27] The amino group is less acidic than the carboxyl group and therefore in an aqueous solution at a pH of between 4 and 9, the amino acid exists as a zwitterion and loses a proton from the amino group.[27,28] Thus, the amino acids are able to form five-membered chelate rings.

"In certain instances resonance effects, tending to increase stability, may be enhanced when the metal ion participates in the resonance of the ligand in the metal chelate ring."[25] This is a fourth factor that influences the stability of the chelate.

A fifth factor that influences chelation stability is the steric effect of the ligand.[25] Certain ligands tend to keep donor groups in certain positions that are most favorable to coordination with the ion. When this occurs, stability is increased. If another ligand is substituted for the first, it may introduce steric effects that in some cases will reduce chelate stability or prevent chelation from occurring at all.[27]

Substitution of a ligand may also alter the basic strength of the ligand. As noted above, ligands act as Lewis bases, whereas the cations act as Lewis acids. The more basic the ligand, the greater its stability. In the case of bidentate ligands, which include the amino acids, the metal ion can be coordinated with two basic groups, one basic and one acidic group, or two acidic groups. As ligands are substituted and these groups change, so does the stability of the chelate.[22,27]

A sixth consideration in discussing the relative stability of a chelate is the nature of the donor atoms.[27] As required by definition, a chelate must contain at least two donor atoms that are capable of bonding the same metal ion. The donor atoms are the more electronegative ones which are located on the right-hand side of the periodic chart primarily in Group V (nitrogen, phosphorus, arsenic, and antimony) and Group VI (oxygen, sulfur, selenium, and tellurium). "Elements below nitrogen in Group V also show donor properties but these are less marked than those of nitrogen-containing ligands. The ability to act as a donor falls off as the atomic number increases."[32] The elements in Group VII, the halogens, behave as non-chelating, unidentate ligands (where only one pair of electrons is involved) in mononuclear complexes. When they are covalently bound, they do not act as donors in chelates.[27,30]

The donor atoms that form part of the basic group ($-NH_2$, -NH, -N=, O, -O-, =N-OH, -OH, -S-, $-PR_2$, and -As, R_2 carry a lone pair of electrons which may interact with the cation. The acidic groups (-COOH, $-SO_3H$, $-PO(OH)_2$, -OH, +N-OH, and -SH) lose a proton and coordinate with the metal ion.[30] Thus, as donor atoms change within a given ligand, the basic strength also changes with a corresponding charge in the stability of the chelate.

Furthermore, since donor atoms vary in their electronegativity,[30] changes in one of them without a ligand also changes its electronegativity. As the electronegativity of a ligand changes, so does the stability of the chelate.[31]

Finally, the metal ion occupying the central position within the chelate molecule also influences its stability. Like the ligand, the electronic structure of the metal ion, its size and oxidation state, its coordination number, and its stereochemistry all influence the stability of the chelate.

Examining the first of these metal ion characteristics, it should be understood that the presence of incomplete filling of electrons in the outer shell of the metal atom confers special properties on the metal. This includes variable valences. Furthermore, chelates of the transition metals, the lanthanides and the actinides, are more stable than corresponding alkali metal chelates.[33]

The size of the ion is the next consideration. Its size depends upon its coordination number, the nature of the molecules or ions it is attached to, and the extent to which the bonding departs from a purely electrostatic attraction. The coordination number is influenced by the charge on the cation. When the charge is low, only a few lone pairs of electrons can become involved. When the charge is higher, such as +3 or +4, a higher coordination number can be reached. Thus, the larger the metal ion, the greater the number of ligand atoms that can surround it and be regarded as being in contact with the central ion.[29] As a general rule the number of bonds is twice the valence of the metal.

The coordination number also affects the stereochemistry of the metal chelate. For example, Table 4-2 presents the changing stereochemistry of various transition metals.[34] As they change shapes, so does the stability of the resulting chelate.

As the above superficial discussion has demonstrated, the relative stability of a chelate is a highly complex subject. Only stability in an aqueous medium has been approached. Both temperature and oxidation and reduction stability also affect the overall stability of the chelate.[35] Temperature fluctuations within the gut are probably not extreme enough to be of concern, and oxidation and reduction are outside the scope of this book. Thus, it is solution stability, the medium within which the chelates found in the gut and the cytoplasm of the mucosal cell exist, that is of great importance to this discussion.

Hughes has summarized some of the solution stability constants for various metals chelated with amino acids in which the central ion assumes different valences.[36] This is shown in Table 4-3.

Table 4-2.

STEREOCHEMISTRY AND OXIDATION STATES OF SOME 3d ELEMENTS					
OXIDATION STATE	COORD. NUMBER	STEREOCHEMISTRY	OXIDATION STATE	COORD. NUMBER	STEREOCHEMISTRY
$Cu^{I}(d^{10})$	2	linear	$Co^{III}(d^{6})$	4	tetrahedral
	3	planar		5	square pyramidal
	4*	tetrahedral		6*	octahedral
$Cu^{II}(d^{9})$	4*	square planar	$Co^{IV}(d^{5})$	6	octahedral
	4	distorted tetrahedral	$Fe^{II}(d^{6})$	4	tetrahedral
	5	square pyramidal		5	?
	5	trigonal bipyramidal		6*	octahedral
	6*	distorted octahedral	$Fe^{III}(d^{5})$	4	tetrahedral
$Cu^{III}(d^{8})$	6	octahedral		6*	octahedral
$Ni^{II}(d^{8})$	4*	square planar		7	pentagonal
	4*	tetrahedral			bipyramidal
	5	trigonal bipyramidal	$Fe^{IV}(d^{4})$	6	octahedral
	6*	octahedral	$Fe^{V}(d^{3})$	4	tetrahedral
$Ni^{III}(d^{7})$	5	tirgonal bipyramidal	$Mn^{I}(d^{6})$	6	octahedral
$Ni^{IV}(d^{6})$	6	octahedral	$Mn^{II}(d^{5})$	4	tetrahedral
$Co^{I}(d^{8})$	4	square planar		4	square planar
	5	trigonal bipyramidal		6*	octahedral
	6	octahedral	$Mn^{III}(d^{4})$	6*	octahedral
$Co^{II}(d^{7})$	4*	tetrahedral	$Mn^{IV}(d^{3})$	6	octahedral
	4	square planar	$Mn^{V}(d^{2})$	4	tetrahedral
	5	trigonal bipyramidal	$Mn^{VI}(d^{1})$	4	tetrahedral
	5	square pyramidal	$Mn^{VII}(d^{0})$	3	planar
	6*	octahedral		4	tetrahedral

*Most common states.

In summary then, in biology, metals are firmly bound to amino acids and complex proteins through the process of chelation. Neither calcium nor magnesium move as freely as simple hydrated ions as they do as chelates when in biological fluids as demonstrated by radioisotope and ion-selective electrodes. "It seems likely that the immediate environment of the metal ion is altered appreciably during its transport through the membrane and quite possibly movement is facilitated by chelation processes involving naturally occurring ligands."[37] And finally, as the following chapters of this book will demonstrate, when the metal ion is pre-chelated with two amino acids to form a dipeptide-like molecule prior to oral ingestion, its mucosal uptake is thought to be facilitated through a second intestinal absorption mechanism which is more efficient than the ionic absorption system that relies on membrane-bound integral proteins to function as transport ligands.

Table 4-3.

LOGARITHMS OF FORMATION CONSTANTS IN AQUEOUS SOLUTION AT 25° C

Ligand	Ionic strength	Constant	Mg^{2+}	Ca^{2+}	Mn^{2+}	Fe^{2+}	Fe^{3+}	Co^{2+}	Ni^{2+}	Cu^{+}	Cu^{2+}	Zn^{2+}
Glycine	→ 0	K_1	3.44	1.38	3.44	4.3	—	5.23	5.77	—	8.62	5.52
	→ 0	K_2	3.01	—	—	3.5	—	4.02	4.80	—	6.97	4.44
Cysteine	0.01	K_1	<4	—	4.1	—	—	9.3	—	19.2	—	9.86
	0.01	K_2	—	—	—	—	—	7.6	—	—	—	8.84
	→ 0	β_2	—	—	—	11.77	32.1	—	19.3	—	—	—
Serine	Varied	K_1	—	~0.5	—	—	—	—	—	—	—	—
	Varied	β_2	—	—	—	7.0	—	8.0	—	—	14.54	—
Arginine	0.15	K_1	—	—	—	—	—	3.87	—	—	—	—
	0.15	K_2	—	—	—	—	—	3.20	—	—	—	—
	0.15	K_3	—	—	—	—	—	2.08	—	—	—	—
	0.01	β_2	—	—	—	—	—	—	—	—	13.90	7.80
Aspartic acid	0.10	K_1	2.43	1.60	3.74	—	—	5.90	7.12	—	8.57	5.84
	0.10	K_2	—	—	—	—	—	4.28	5.27	—	6.78	4.31
Histidine	0.01	β_2	—	—	7.74	9.3	—	13.86	15.9	—	18.33	12.88
Lysine	0.01	K_1	—	—	2.0	4.5	—	—	—	—	—	—
	0.01	β_2	—	—	—	—	—	6.8	8.8	—	13.7	7.6
Glutamic acid	Varied	K_1	1.9	2.05	3.3	4.6	—	5.06	5.90	—	7.85	5.45
	—	K_2	—	—	—	—	—	3.40	4.44	—	6.55	4.01
Glutamine	0.16	K_1	—	0.18	—	—	—	—	—	—	—	—
Asparagine	0.01	β_2	~4.0	—	~4.5	6.5	—	8.13	10.6	—	14.9	8.7
Glycylglycine	→ 0	K_1	1.06	1.24	2.15	—	—	3.49	—	—	6.04	3.80
		K_2	—	—	—	—	—	2.39	—	—	5.62	2.77
Glycylserine	0.01	K_1	—	—	—	—	—	—	—	—	—	3.7
Alanylglycine	0.01	K_1	—	0.66	—	—	—	—	—	—	—	3.0
Glycylalanine	0.01	K_1	—	—	—	—	—	3.15	—	—	—	4.1
	0.01	K_2	—	—	—	—	—	2.58	—	—	—	—
Glycylglycylglycine	→ 0	K_1	—	—	1.41	—	—	2.98	—	—	5.41	3.33
	—	K_2	—	—	—	—	—	2.61	—	—	5.15	2.99
Imidazole	Varied	K_1	—	0.08	—	—	—	—	2.94	—	4.20	2.58
		K_2	—	—	—	—	—	—	2.41	—	3.47	2.19
		K_3	—	—	—	—	—	—	1.99	—	2.84	2.41
		β_2	—	—	—	—	—	—	—	10.87	—	—
Proline	0.03	β_2	<4	—	5.5	8.3	—	9.3	—	—	16.8	10.2
1,10 Phenanthroline	Varied	K_1	—	—	—	—	—	—	—	—	6.30	—
	—	K_2	—	—	—	—	—	—	—	—	6.15	—
	—	K_3	—	—	—	—	—	—	—	—	5.50	—
	—	β_3	—	—	7.35	21.0	14.10	—	—	—	—	—
2-methyl.1.10 phenanthroline	—	β_3	—	—	—	10.8	—	—	—	—	—	—
Ammonia	—	K_1	0.23	0.2	—	—	—	2.11	2.79	5.93	4.15	2.37
	—	K_2	0.15	0.6	—	—	—	1.63	2.24	4.93	3.50	2.44
	—	K_3	0.42	0.8	—	—	—	1.05	1.73	—	2.89	2.50
Ethylenediamine	1-0M	K_1	0.37	—	2.73	4.28	—	5.93	7.66	—	10.72	5.92
	—	K_2	—	—	2.06	3.25	—	4.73	6.40	—	9.31	5.15
	—	K_3	—	—	0.88	1.99	—	3.30	4.55	—	—	1.86
1.3 diaminopropane	0.15	K_1	—	—	—	—	—	—	6.98	—	9.77	—
	—	K_2	—	—	—	—	—	—	4.93	—	7.17	—
Ethanolamine	→ 0	β_2	—	—	—	—	—	—	—	—	6.68	—
2-mercapto ethylamine	1.0	K_1	—	—	—	—	—	9.38	10.96	—	—	8.07

REFERENCES

1. Craig, D. and Nyholm, E., "Nature of the Metal Ligand Bond," in Dwyer, F. and Mellor, D., eds., *Chelating Agents and Metal Chelates* (New York: Academic Press), 53, 1964.
2. Bailar, J. and Busch, D., "A General Survey of the Coordination Compounds," in Bailar, J., ed., *The Chemistry of Coordination Compounds* (New York: Reinhold), 1-99, 1956.
3. Bell, C., *Metal Chelation Principles and Applications* (Oxford: Clarendon Press), 1, 1977.
4. Baxter, J. and Steiner, L., *Modern Chemistry* (Englewood Cliffs: Prentice Hall), VI, 1959.
5. Chenoweth, M., "Physicians Synopsis of Chelation Chemistry," in Soffer, A., ed., *Chelation Therapy* (Springfield: Charles C Thomas), 146, 1964.
6. Craig, op. cit., 52
7. Mellor, D., "Historical Background and Fundamental Concepts," in Dwyer, F. and Mellor, D., eds., *Chelating Agents and Metal Chelates* (New York: Academic Press), 17, 1964.
8. Bell, op. cit., 9.
9. Mellor, op. cit, 3.
10. Ibid., 28.
11. Ibid., 20, 44.
12. Martell, A., "The Relationship of Chemical Structure to Metal-Binding Action," in Seven, M., ed., *Metal Binding in Medicine* (Philadelphia: J.B. Lippincott Co.), 17, 1960.
13. Barclay, G. and Hoskins, B., "The Crystal Structure of Carbontotetra-Amminecobalt (III) Bromide," J. Chem. Soc. Proc., 586, 1962.
14. Bell, op. cit., 27-32.
15. Ibid., 12.
16. Ibid., 16, 17-18.
17. Wallace, A., "Metal Chelation and Mechanisms of Metal Function in Biological Systems," in Wallace, A., ed., *A Decade of Synthetic Chelating Agents in Organic Plant Nutrition* (Los Angeles: Arthur Wallace), 7, 1962.
18. Furst, A., *Chemistry of Chelation in Cancer* (Springfield: Charles C Thomas), 12, 1963.
19. Morgan, G. and Drew, H., "Residual Affinity and Coordination. II. Acetylacetones of Selenium and Tellurium," J. Chem. Soc., 117:1456, 1920.
20. Moeller, T. and Horwitz, E., "Chelation," in Comar, C. and Bronner, F., eds., *Mineral Metabolism* (New York: Academic Press), VI, 109, 1960.
21. Wallace, op. cit., 9.
22. Bell, op. cit., 11.
23. Eichhorn, G., "Introduction to Chelation Chemistry," in Soffer, A., *Chelation Therapy* (Springfield: Charles C Thomas), 135, 1964.
24. Craig, op. cit., 41.
25. Martell, op. cit., 2.
26. Bell, op. cit., 17.

27. Mellor, op. cit., 45.
28. Hughes, M., *The Inorganic Chemistry of Biological Processes* (London: John Wiley & Sons), 2-4, 1972.
29. Bell, op. cit., 32-34.
30. Ibid., 9.
31. Mellor, op. cit., 21.
32. Bell, op. cit., 14.
33. Hughes, op. cit., 25-26.
34. Mellor, op. cit., 40.
35. Hughes, op. cit., 71-72.
36. Bell, op. cit., 107.
37. Werner, A., *Neuere Anschauungen auf dem Gebiete der Anorganischen Chemic*, (Brunswick: F. Vieweg & Sohn), 1920.
38. Ley, H., Z. Elektrochem, 10:954, 1904.

CHAPTER 5

INTESTINAL UPTAKE OF FREE METAL IONS

ALTHOUGH there has been a voluminous amount of published discussion covering mineral absorption from the intestines, actual definitive descriptions are rare. In many instances for certain metals they are nonexistent. This is due in part to inadequate instrumentation that allows one to see the actual absorption of an ion through the cellular membrane. Consequently, the theories proposed for an absorption model are based on indirect observations.

Because of these problems most researchers have limited themselves to the question of whether or not a specific mineral is absorbed through a specific region in the intestine. Having ascertained that, they proceed to determine the effects that specific additives or deletions to the intestinal environment have on mineral uptake.

These studies have been most useful. In the course of this research certain repetitive patterns have evolved for several minerals which strongly suggest a generally common type of intestinal mechanism or absorption pathway for the divalent cations under consideration (calcium, magnesium, zinc, iron, copper, and manganese).

A discussion of ionic mineral absorption from the intestine is best prefaced by briefly reviewing the barriers the metal must traverse in its path from the lumen to the blood. Using Figure 5-1 as a reference, "The luminal side of the intestinal absorptive cell is covered with a striated border, the microvilli. The microvilli are enclosed by a plasma membrane, a typical trilaminar membrane consisting of electron-dense outer and inner layers covering an electron-lucent

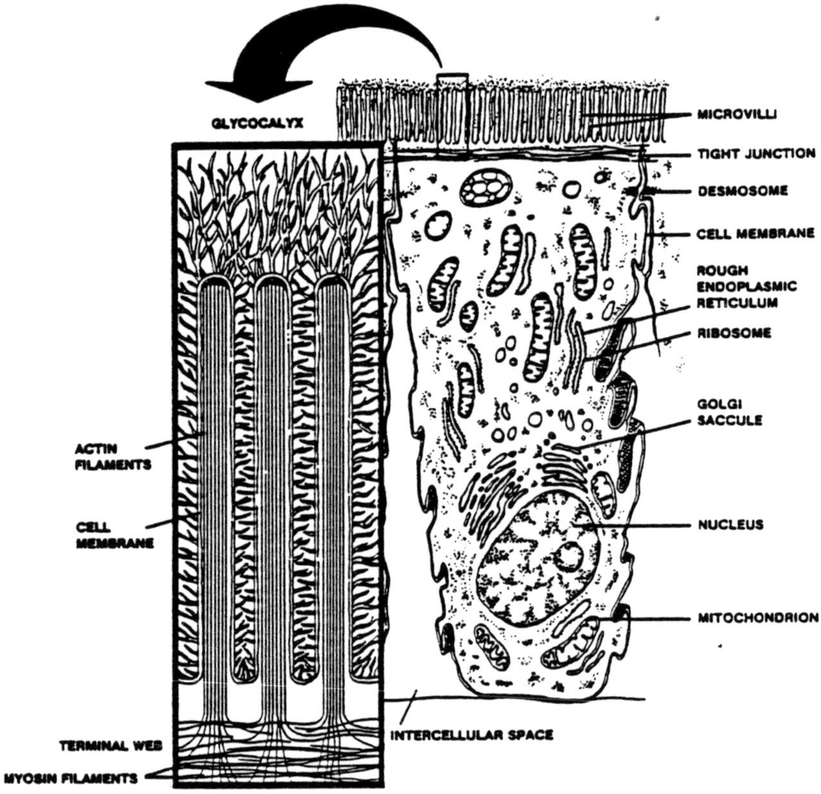

Figure 5-1. Schematic diagram of a mucosal cell showing the microvilli.

central layer.

"The core of the microvillus contains bundles of filaments that run parallel to the long axis of the microvilli and intermesh with a network of filaments and tubules in the apical cytoplasm of the absorptive cell. The cytoplasm consists of the usual cellular array of organelles, including lysosomes, mitochondria, endoplasmic reticulum, Golgi material, and a nucleus located at the basal portion of the cell.

"The basolateral plasma membrane, like the apical plasma membrane, has a trilaminar structure but differs from the apical membrane in size and enzyme content. The basolateral membrane is in direct contact with the intercellular fluid, but this fluid is sealed from the gut lumen by a 'tight junction' at the apical end of the absorptive cells. Thus, an ingested nutrient must traverse the apical plasma membrane, the cytoplasm, and the basolateral plasma membrane before reaching the intercellular space.

"After crossing the cytoplasm and membranes of the cell, nutrients must pass through the basement membrane, a homogeneous layer of carbohydrate-protein complexes that separates the bases of the intestinal cells from the lamina propria. In the lamina propria, absorbed nutrients enter the blood either directly by passing through the capillary walls or indirectly by entering the lacteals, which deliver lymph to the blood via the thoracic duct.

"Considering the morphological barriers confronting a zinc ion during its course from the lumen to the blood, the likelihood of simple diffusion taking place seems remote. The absorption of zinc, and apparently of other metals, is probably regulated or at least facilitated by small ligands and/or metal-binding proteins."[1] This chapter discusses investigations into mineral absorption conducted at the cellular and the molecular level.

If the mineral salt is a soluble salt, it presumably will ionize in the acid pH of the stomach after ingestion. For ionic transport through the mucosal cells of the intestine it is essential that the mineral ionize prior to entering the duodenum which is the first ten to twelve inches (25 to 27 cm.) of small intestine in man.[2] This area of the small intestine has about the same pH as the stomach (approximately a pH of 2)[3] and tends to help maintain the solubility of the mineral salt.

Investigators have reported that in the form of an inorganic salt every one of the minerals under consideration must be solubilized prior to absorption. Solubilizing the mineral salt results in a cation. "The rate of absorption of metal ions can be influenced by the pH of the intestinal lumen. The pH may affect the mucosal surfaces or the availability of the luminal contents by altering solubility characteristics, or both effects may operate simultaneously. In general, the more alkaline the lumen, the lower the rate of absorption."[4]

To illustrate, Wasserman and Taylor[5] have reported that the most efficient absorption of calcium in terms of moles of calcium per unit time per gut length occurred in the low pH found in the duodenum. Nevertheless, they hasten to add that calcium is absorbed from all sites of the intestinal tract. It is true there is generally more absorption in the ileum due to the longer resident time in that segment,[6] but since the ileum has a higher pH than the duodenum, solubility of calcium ions will usually decrease[7] resulting in lower absorptive efficiency. Radioassays have shown that the intestinal capacity to absorb ^{47}calcium decreases in the following order: Duodenum-jejunum-ileum.[8] This is seen in Figure 5-2. Calcium is more soluble in an acid medium and hence more readily absorbed from the duodenum.[9,10]

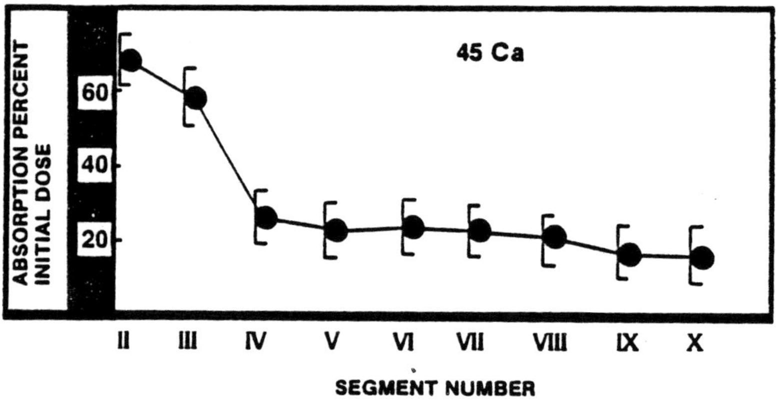

Figure 5-2. The rates of absorption of Ca^{2+}. Solutions containing ^{45}Ca salts were introduced into the individual segments of rats' intestinal lumen, ligated into nine equal segments. Segment II includes the duodenum, from the pylorus to the ligament of Trietz. Segments were excised after 30 minutes; residual activity was assayed in a gas flow β-counter. Percentage of initial dose absorbed is recorded. Each point represents mean $\pm$ S.D. (12 rats).

Copper is also more rapidly absorbed from the intestine when the luminal contents are still acidic.[11,12] In writing about copper toxicity, Scheinberg and Sternlieb have reported that when copper salts, such as copper sulfate, come in prolonged contact with acidic food or drink (vinegar, carbonated beverages, citrus juices, etc.), acute copper poisoning can occur[13] thus pointing out the high solubility of the copper ions in an acidic pH. Research with intestinal sections removed from pigs has shown the proximal duodenum to be the most efficient site of uptake for copper ions,[14] presumably because of its acidic environment. Indeed not only is copper absorbed from the upper small intestine, but also a small amount can be absorbed from the stomach.[15,16,17]

The pH of the intestines also influences the absorption of magnesium from $MgSO_4$, magnesium aspartate, magnesium aspartate hydrochloride, and $MgCl_2$ conducted in Sprague-Dawley male rats *in vivo*. The magnesium compounds were dissolved in10 ml/Kg of demineralized water and administered twice daily by lavage. On the third day the animals were sacrificed and their plasmas and femurs assayed by atomic absorption spectrophotometry.

The investigators found, "in normal rats, oral administration of $MgSO_4$ does not influence Mg levels in plasma and bone. Magnesium aspartate at a dose of 250 mg/Kg increases plasma Mg; bone Mg is significantly increased only when the dose is doubled to 500 mg/Kg. In addition to hypermagnesemia, the two Cl-containing compounds (Mg asp-HCl and $MgCl_2$) cause significantly increased Mg concentrations in bone at a dose of 250 mg/Kg ($P<0.001$ compared to controls) and are superior to $MgSO_4$ and Mg asp . . . Thus it appears that Mg uptake into plasma and bone is significantly improved by Cl".[18]

The role of the chloride ions in this experiment undoubtedly was to lower the pH of the intestine with corresponding increase in magnesium absorption. This is consistent with findings by Chutkow who reported that the duodenum is the most active site for magnesium absorption.[19] Magnesium salts are poorly absorbed in alkaline pH.[20]

Iron is only soluble in the acidic medium of the duodenum according to Forth and Rummel.[21] Consequently, greatest absorption of it occurs in the acid pH environment of the upper portion of the small intestine. When ^{59}Fe was administered orally to rats and then their small intestines removed, washed, cut into segments and the retained radioactivity measured, the gradient shown in Figure 5-3 was developed. As is seen, maximum activity was in the proximal duodenum with very little activity in the ileum.[22]

For absorption of iron from ferrous or ferric salts to occur, the iron must first be ionized. Any condition which causes the formation of insoluble iron salts will interfere with iron absorption.[23] For this reason the inclusion of such additives as ascorbic acid which lower the intestinal pH and keep the iron salts soluble, has long been regarded as an important factor in iron absorption. When the pH of the gut is lowered, absorption is enhanced.[24]

The pH of the intestine also influences the absorption of zinc.[25] Reinhold, *et al.* have reported, "The binding of zinc . . . is dependent upon pH. Solubility of zinc is minimal in the vicinity of pH 7.0. It

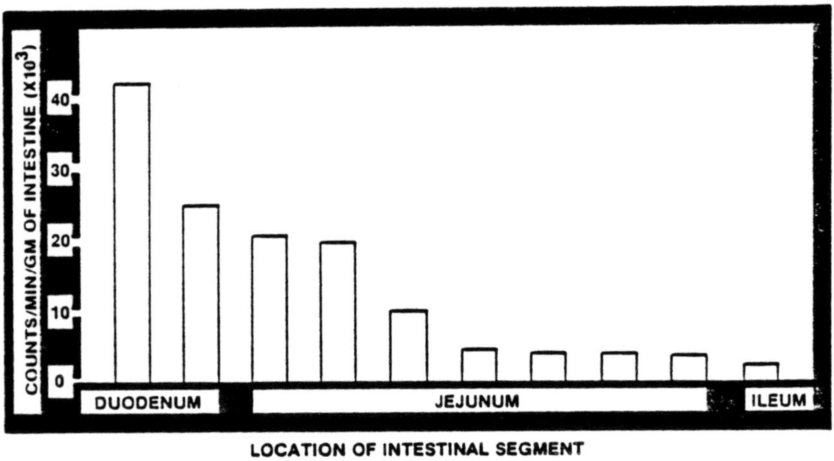

Figure 5-3. The absorption of radioactive iron at different sites of the intestinal tract of the rat.

rises appreciably as the pH approaches 6.0, then more rapidly in more acid solution."[26] When research animals were fed ^{65}Zn, it was found that the intestinal tract absorbs zinc at different rates. The duodenum has the highest rate of absorption where the pH is the lowest. It is followed by the ileum and jejunum.[27,28]

Manganese is soluble only in the acidic pH of the duodenum.[21] The precise loci in the gastrointestinal tract for the absorption of manganese have not been identified,[29] but it is known that iron competes with manganese for absorption sites in the lumen.[30] That suggests that the duodenum is the most logical site for the majority of the absorption of the manganese ion because it is only there in the acidic pH that solubility is assured. In emission spectrographic analysis for manganese in rat intestine, more than twice as much manganese was absorbed from the duodenum as from the jejunum and almost twice as much as from the ileum. The mean absorption of manganese from the duodenum almost equalled the combined mean absorption from the jejunum and the ileum.[31]

Once the mineral salt has been ionized in the gastric acid of the stomach, it presumably remains soluble in the upper intestine before the introduction of the bile and neutralizing pancreatic buffers and enzymes occurs. The cation is electrically attracted to the protein molecules in the mucosal cell wall membrane. The average number of moles of a cation bond increases at more acid pH and with increase in net positive charge on the protein.[32,33,34]

The pH of the intestine raises as the chyme descends the small intestine and is mixed with the alkaline secretions from the pancreatic and bile ducts.[21]

The bile is secreted by the gallbladder. It is originally produced in the liver. The bile passes from the liver to the gallbladder where it is stored and concentrated until it is required. Upon stimulation of chyme entering the duodenum and the mucosal hormone choleciptekimi, the bile is released where it converts the fats and oils into a milky liquid in preparation for their absorption. The pH of the bile from the gallbladder is 6.5[35] Although still acidic, it helps raise the pH from that of 2 in the stomach towards the alkaline pH.

Through the same papilli the pancreatic fluid enters the duodenum. It is the only digestive fluid that digests starches, carbohydrates, fats and proteins. It is highly alkaline, deriving its buffering capacity from the blood and to a small extent from the CO_2 produced by the pancreatic cells.[36] The high bicarbonate content of the pancreatic juice helps solubilize nutrients to a concentration that is optimum for digestion and absorption. "Its high bicarbonate content helps to neutralize acidic gastric chyme emptied into the duodenum and to maintain pH in the optimal range for most pancreatic enzymes to act. A key aspect of this neutralizing function is the requirement that at the level of the papilla of Vater and beyond, intraluminal pH should not decrease below 4.0; otherwise, lipase, trypsin and other pancreatic enzymes may be irreversibly inactivated. . ."[37]

The maintenance of this higher pH is also assisted by the secretion of an alkaline fluid by glands embedded in the mucosal lining of the small intestine. It should be remembered that the first steps of digestion of food in the stomach requires pepsin followed by pancreatic protease which together break some of the protein bonds. Pepsin can function only in a low pH environment (high acid). Thus, hydrochloric acid is secreted in the stomach.[38] Consequential to this lowering of the pH is the ionizing of soluble mineral salts. But then as a result of the necessity to raise the intestinal pH so that the pancreatic digestive enzymes can work, the mineral ions reprecipitate. Consequently, if the mineral ion is going to be absorbed from the intestine, it must be done prior to the raising of the intestinal pH. Figure 5-4 illustrates this concept well by showing that calcium, magnesium, iron, and presumably the other bivalent metal ions, are absorbed for the most part prior to passing the common bile and pancreatic papilli.[3]

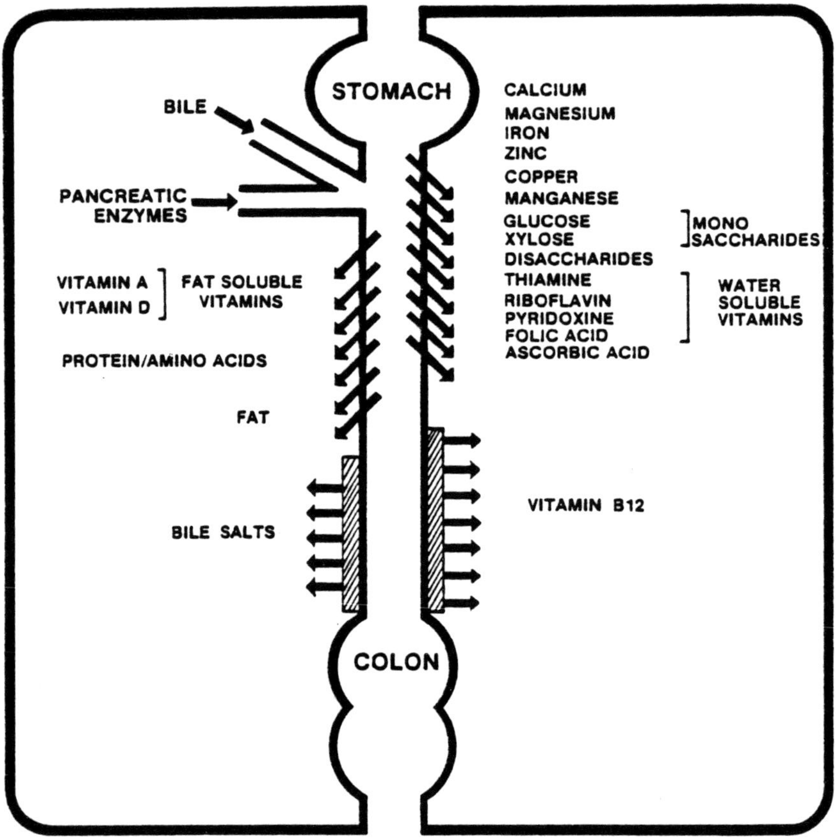

Figure 5-4. Absorption sites on the small intestine for nutrients.

The absorption of calcium occurs in the small intestine[39] and specifically in the upper portion of the small intestine—that is, the duodenum (Figure 5-4).[40] Magnesium is also absorbed from the same area of the small intestine.[3,41] Most iron absorption is localized in the duodenum and the jejunum where there is much higher absorption activity than in the ileum.[3,42] Ionic zinc has been reported to be absorbed in the duodenum of the small intestine.[43] Likewise, manganese is also bound and absorbed in the duodenum.[44] Copper, too, is absorbed from the acid pH environment of the upper portion of the gastrointestinal tract.[45]

Nevertheless, because the hydrogen ion concentration decreases as the ion moves down the duodenum towards the bile duct through the process of chelation, the ion must either be attached to the mucosal wall by the integral binding proteins in the mucosal cell membranes very quickly or the mineral will no longer be as available for absorption. This ion-binding protein is a determining factor for the absorptive utilization.[46] It is important that all of the mineral ions be capable of binding to protein in the gut because this binding (chelation and/or complexing) is an essential step in the intestinal uptake of the metal ion.

The chelation of a polyvalent ion by an integral protein in the membrane requires that it furnish two negative donor groups to combine with the metal. One of the donor groups generally comes from the amino group (NH_2) and forms a complex covalent bond with the metal.[47] The other donor group comes from the carboxyl group (COOH).[48] The resulting structure is a heterocyclic ring as shown in Figure 5-5 which contains at least five bonds.[49] This concept was discussed in greater detail in Chapter 4.

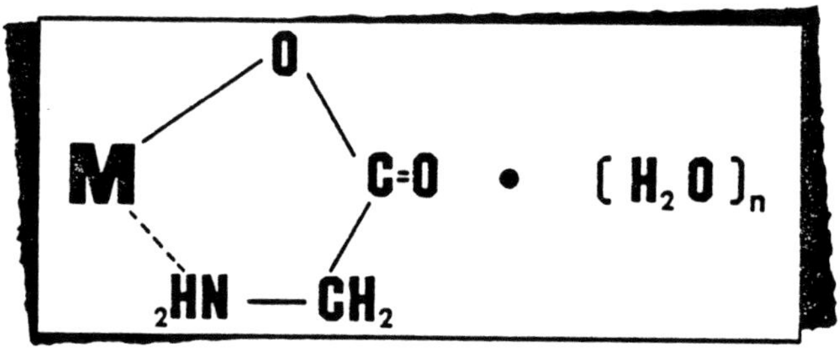

Figure 5-5. An amino acid chelate where M represents the metal.

Wasserman, *et al.*, discovered a calcium-binding protein in the mucosal cell which is believed to be one of the calcium transport components.[50] It is dependent upon 1, 25-dihydroxyvitamin D_3 for its production.[51] As seen in Figure 5-6,[52] it is believed that 1, 25-$(OH)_2D_3$ reacts with a 3.75 cystol receptor. Together they are transferred to the nucleus of the mucosal cell where the vitamin is bound to specific sites on the chromatin. This in turn leads to a transcription of messenger RNA for the synthesis of the calcium transport protein.[52]

This vitamin D dependent calcium transport protein was ob-

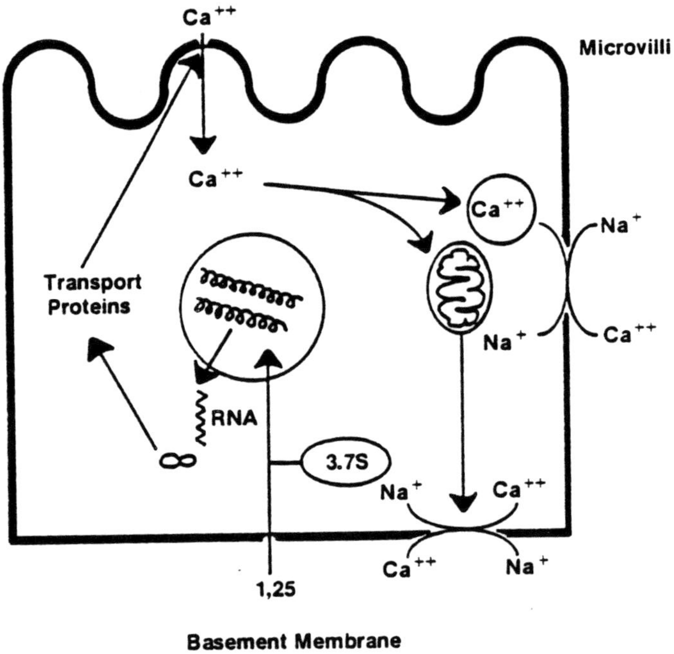

Figure 5-6. Diagrammatic representation of the postulated mechanism whereby 1,25-$(OH)_2D_3$ stimulates intestinal transport of calcium in the intestinal epithelial cell.[52]

tained from cells with a high degree of purity in 1968[53] through a three-step process: (1) ammonium sulfate (75% saturated) fractionation; (2) Sephadex G-100 gel filtration; and (3) preparative discontinuous electrophoresis. About 2.5% to 3% of the protein in the cell is a calcium-binding protein. It has a molecular weight of between 24,000 and 28,000. It is not a lipoprotein and contains little, if any, carbohydrates, phosphates, or sialic acid. It does contain a high concentration of glutamic acid, aspartic acid, leucine, and lysine.[54]

The sediment coefficient is about 2.96 Svedberg units at 20°C.[53]

The formation constant of the resulting calcium-protein chelate was found to be $2.6 \times 10^5 M^{-1}$.[53] This is about 100 times greater than a calcium-citrate chelate,[55] about 1000 times stronger than a calcium-albumin complex,[56] and about the same strength as the formation constants for calcium-binding protein in the muscle.[57]

The calcium-binding protein in the mucosal cell has the capacity to bind one mole of calcium per mole of protein.[53] It is found in all parts of the small intestine with a higher concentration in the duodenum, followed by the jejunum, and finally the ileum.[58] Calcium absorption decreases in that same order.[54,58,59]

As stated earlier, vitamin D appears to be a prerequisite for the formation of the calcium-binding protein in the mucosal cell. Zull, *et al.*,[60] have suggested that the vitamin D may first interact with the cell membrane and change its permeability to the calcium. This may alter the intranuclear calcium concentration. Those events lead to the formation of the calcium-binding protein which then chelates and translocates the calcium as suggested in Figure 5-6.

Using fluorescent antibody tracing methods, Taylor[61] discovered that another calcium-binding protein (CaBP) is located in glycocalyx and/or the brush border of the microvilli. "It is not unreasonable to assume that functional calcium-binding protein is that present along the microvillar border, and the goblet cell represents the site of synthesis or localization of the protein. The sensitivity of the procedure is insufficient to indicate if CaBP is located only on the microvillar border or within the cell proper. However, in the former location the protein is in a position to interact with luminal Ca as one of the first steps in the absorptive process."[62]

The calcium-binding protein can interact with lysolecithin to form a complex which does not dissociate in an electrical field. It is also known that the chelating activity of the protein is partially and rapidly inhibited by a phospholipid. These findings aid in formulating a model that includes these functions. First the calcium-binding protein penetrates the matrix of the lipid membrane of the epithelial mucosal cell where it is identified as an integral protein in Singer's fluid mosaic membrane model. The ionic calcium is chelated by this protein at the interface of the luminal fluid and microvillar membrane. The calcium is released within the hydrophobic region of the membrane. This is due to the observations mentioned above. This release then allows the calcium to diffuse across the balance of the membrane into the cytoplasm or back into the luminal fluid.[63]

Zinc is also carried from the lumen into the mucosal cells by a protein carrier.[64] The exact chemical nature of the zinc-binding ligand is not clear. Hahn *et al.*, isolated a ligand from the duodenum of a pig. After it was purified, its structure was determined by nuclear magnetic resonance (NMR). The ligand was tentatively identified as N,N,N-trimethyl-1, 2-ethandiamino, which has a molecular weight of 800. Other researchers found that the binding ligand was a peptide.[66] Song and Adham[67] identified another zinc-binding ligand which was determined to be a prostaglandin.[68]

In later research Schricker[69] isolated this zinc-binding ligand and submitted it to extensive chemical analysis. He found that it is neither N,N,N-trimethyl-1, 2-ethandiamino nor prostaglandin. Instead it is, as originally reported by Evans, *et al.*, a peptide. Because a similar ligand was extractable from pancreatic secretions, Evans proposed that this zinc-binding protein is secreted into the lumen from the pancreas where it chelates the luminal zinc and carries it into the mucosal cells.[43]

Perhaps that is not exactly correct because Schricker recovered the ligand from the mucosal cells of the intestine rather than from the lumen, and none of his experiments proves or even indicates that the binding protein actually originated in the pancreas (although he agrees with Evans' model). What Schricker does prove is that the zinc-binding ligand is a peptide and it is found in the mucosal cells.

There is a large variety of amino acids present in this ligand. However, aspartic acid, glutamic acid, glycine, cysteine, and lysine in a molar ratio of 2:3:3:1:1, respectively, are predominant. The total molecular weight of the binding ligand is less than 1,500 daltons.

The sulfhydryl group of the cysteine has been postulated as being the most probable position to bind the zinc ion formed in the gut. This results in the formation of a chelate similar to the one represented in Figure 5-7.

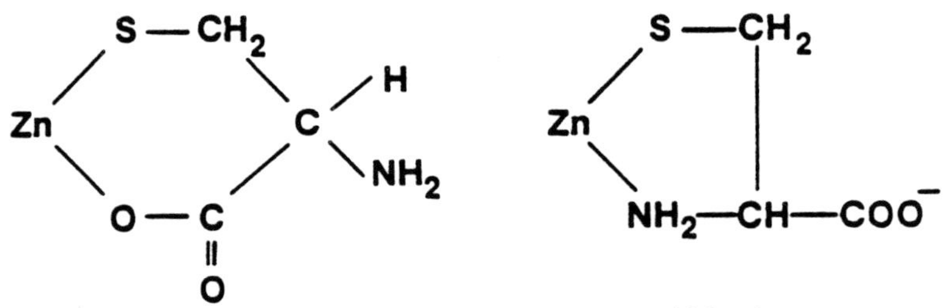

Figure 5-7. Chelation of Zinc in the Mucosal Membrane.

This binding ligand has a bipolar nature which would be advantageous for a molecule formed in the cytoplasm of the mucosal cell but must pass through the lipid cellular membrane to bind the zinc in the gut. Schricker postulated that aspartic acid or asparagine and the glutamic acid or glutamine may account for the charge properties of the ligand.

The location of this binding ligand has been suggested by the research of Evans, Grace, and Votava.[66] In orally administered ^{65}Zn, 12% was associated with the microvillus membrane fraction of the small intestine of the rat. Thirty percent of the ^{65}Zn was recovered in the basolateral plasma membrane. Furthermore, when plasma membranes containing a known quantity of ^{65}Zn were incubated in a medium which contained albumin, 96% of the ^{65}Zn was transferred to the medium while less than 30% of the zinc was released to other media that contained a 3 to 1 zinc-albumin complex or no albumin at all. These results suggest that the zinc-binding ligand is located in the basolateral plasma membrane and is thus able to mediate the absorption of zinc.

There is, however, more than one zinc-binding ligand in the mucosa. Rats were given orally ^{65}Zn. Afterwards their small intestines were removed and fractionated. Using a Bio-Gel P-10 column, four isotope peaks were obtained. Two peaks jointly contained 50% of the recovered isotope. The molecular weight of those two ligands was approximately 20,000 daltons. The third fraction had a molecular weight of approximately 10,000 and contained 10% of the recovered zinc. The last fraction had a molecular weight of about 800 and contained the remainder of the recovered zinc.[70] Others have also observed zinc-binding ligands of varying molecular weights in the duodenum.[71]

Due to the work with the transfer of zinc into an albumin medium mentioned above,[66] Evans, *et al.*, proposed that the zinc-binding ligands in the cytoplasm of the mucosal cell are albumin in nature. This is seen in Table 5-1.[39]

This albumin protein is generally called metallothionein and has an average molecular weight of about 10,000 daltons. (In man its molecular weight may be 65,000.) It has a high sulfhydryl content and presumably chelates as seen in Figure 5-7.[72] It is believed to pick up the zinc that has been deposited in the basolateral plasma membrane by the zinc-binding ligand in the membrane and then carry the zinc to the basement membrane of the mucosal cell. Further, it is postulated that this albumin regulates the flow of the zinc into the mucosal cell by the quantity of free albumin available to chelate the

Table 5-1 [39]

EFFECT OF ALBUMIN ON ^{65}ZN REMOVAL FROM ISOLATED EPITHELIAL CELLS OF RATS			
	CELLS	MEDIUM	
ADDITIONS TO MEDIUM	(CPM) [a]	(CPM) [a]	% TOTAL [b]
None	3290	490	13 ± 3
Metal-free albumin	1110	220	67 ± 8
Zn:albumin (1:1)	2740	1330	33 ± 4
Zn:albumin (2:1)	2450	1260	34 ± 5
Histidine, 1 m*M*	8676	172	2 ± 0.5

a-Values shown were taken from one representative experiment.
b-Values shown are the mean SD of four experiments.

zinc in the basolateral plasma membrane.[28] Interestingly enough, some research suggests that vitamin D may have a role in the cellular synthesis of this metallothionein which would make it similar to that of the calcium-binding ligand,[73,74] but the findings are far from conclusive.[59] Its synthesis may be regulated by plasma zinc concentration.[75,76] It is known that mucosal cells synthesize poly (A)-containing RNA (mRNA) which codes for metallothionein.[77] The most important point to remember is that none of the transport of zinc from the lumen to the blood will occur if the ingested zinc salt is not first ionized in the gut. Intact metal salts are not absorbed.

Although not precisely defined, the intestinal uptake and transport of magnesium is related to calcium.[78] For example, Heaton, *et al.*[79] found that the fecal content of magnesium was reduced in four of the six subjects receiving vitamin D_2 (dehydrolachysterol) and Brickman, *et al.*[80] reported that when vitamin D_3 (1,25 dihydroxy-vitamin D_3) was given to three patients, one had a significant decrease in fecal magnesium. Furthermore, there is evidence of the existence of a common mechanism for transporting both calcium and magnesium across the intestinal cells.[81,82]

The first step in the absorption of a salt of magnesium is to ionize it. Its rate of absorption depends to a large extent on the form ingested. The greater the solubility and consequently the dissociation of the cation from the anion, the greater the absorption.[83] "Rat magnesium seems to depend on the amount of endogenous ion present," and "is impaired by luminal substances which form insoluble compounds."[84]

"On the assumption that the [mucosal] cells are impermeable to magnesium ions, it has been suggested that magnesium combines reversibly with a mobile carrier within the outer limits of the intestinal epithelium such as the integral proteins and is transported by it into the epithelial cells. At this point the magnesium ions are released and diffuse into the interstitial fluid. It is assumed that the amount of carrier is limited, thereby exerting a physiologic control over magnesium absorption."[85]

The description of Martin Tansy[85] rings of familiarity. Vitamin D is involved in the production of a carrier protein for calcium. It may be similarly involved in the absorption of the magnesium ion. There is a calcium-binding protein in the border of the microvilli which has been suggested to be an integral protein. It appears that the magnesium cation is also chelated by a protein molecule located in the border of the microvillus. The calcium is probably released within the hydrophobic region of the membrane. There it is picked up by another carrier and transported through the cell. It appears magnesium follows a similar two-step uptake. Thus, it is logical to conclude that both of these alkaline cations follow a similar pathway.

Because the membrane of the mucosal cells is principally lipid, it impedes the direct flow of the ions in and out of the cell. Consequently, when the iron salt is ingested and then ionized in the stomach, the charge on the ion must be neutralized. This is accomplished as proteins embedded in the mucosal cell membrane chelate the iron ions. Presumably this is observed as the binding of the ions to the cell membrane. Freeman and Geer[86] have shown this as probably occurring in Figure 5-8.

The arrow at the microvilli indicates a mass of ferrous carbonate and presumably the binding of the iron to the cell membrane of the microvilli. The arrows in the cytoplasm show iron bound to intracellular ferritin.

As the iron ions are chelated by the proteins located in the muco-

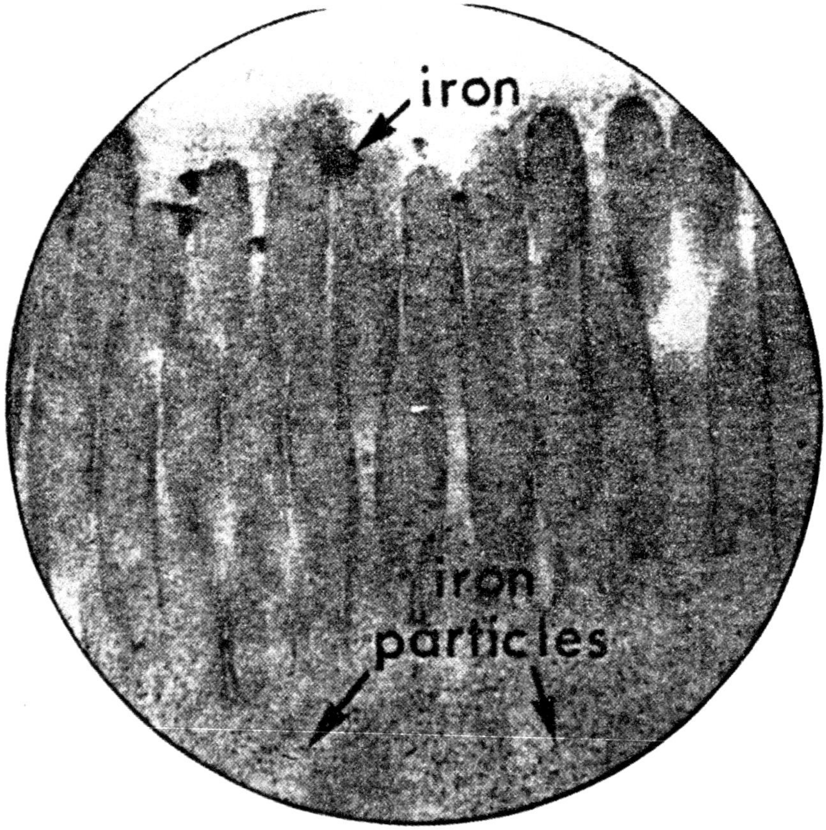

Figure 5-8. The binding of ferrous carbonate on cell membrane.[86]

sal cell membrane, the membrane transport of the iron commences. Facilitated, or carrier mediated, diffusion carries some of the iron across the membrane in an attempt to balance the concentration of iron on both sides of the membrane.[87] Active transport, using a sodium-potassium like pump, is also utilized.[88] It has been theorized that in a manner which is not yet clear, the active transport pump is "turned on" as the bonded minerals begin to diffuse across the membrane. The pump stops operating when there is no more iron bound to the membranous proteins. This will be discussed in greater detail in Chapter 6.

As the iron traverses the membrane and enters the cytoplasm, it is reduced to the divalent state.[89] Perhaps it is this change in form that induces the integral "carrier" protein to release the iron ion to another protein molecule, apoferritin, in exchange for a lipid molecule from the membrane. Iron, like zinc,[66] manganese,[90] copper,[91] calcium,[92] and magnesium[85] requires a two-step absorption mechanism to move from the lumen to the plasma. The first step utilizes a carrier protein that transports the ion across the membrane. This is believed to be the function of the integral protein in the mucosal cell membrane. Once inside the cell the ion is released, presumably due to oxidation and reduction reactions possibly catalyzed by the change in the external and internal pH environment of the cell. It is theorized that the membrane-carrier protein picks up a lipid molecule in exchange for the iron ion.

The second step is the attachment of the ion to another protein molecule of larger size[86] within the cytoplasm itself. The larger ligand molecule would retard the backward diffusion of the mineral because it is now part of a large polypeptide, which is too large to easily traverse the mucosal cell membrane. The formation of a stronger stability constant within the cell due to a different pH may be a further inducement for the iron ion to leave the integral membrane protein in favor of the cell interior protein.[93,94] After the iron combines with the larger protein there is no longer any attraction between the membrane transport protein and the iron ion. Thus, the transport protein is free to be reattached to the lipids in the membrane and to return to its original peripheral protein position as a carrier within the membrane.

Returning to the second step, the iron is bound to apoferritin immediately beneath the plasma membrane,[86] as shown by the small arrow in Figure 5-8. Several investigators now believe the apoferritin, the ligand that chelates the iron, is the same amino acid group as the transmanganin of manganese, copper-thionein, the transport ligand of copper and metallotheonein of zinc.[43]

Apoferritin is a homogeneous globulin with a molecular weight of about 460,000. The central nucleus of the molecule contains the iron which is "stored in six micelles arranged at the corners of a regular octahedron and surrounded by a shell of protein approximately spherical in shape."[95] In the crystalline state it can contain up to 20% iron. It is soluble,[95] which is important in the uptake and release of the iron.

"It is suspected that apoferritin is merely an acceptor of iron for storage purposes. With continued iron loading, larger and more membrane-bound masses of ferritin appear in the apical cytoplasm, thus supporting the concept of storage rather than an absorption regulatory function for apoferritin."[86] Figure 5-9 clearly illustrates this storage concept.[96]

There is a significant gap in our knowledge concerning the movement of the iron ferritin through the rest of the cell to the basement membrane. Hypothetically, the iron may move as a result of osmotic influx and/or diffusion. Moore and Dubach proposed a diffusion model in 1962[97] to explain why the iron leaves the mucosal

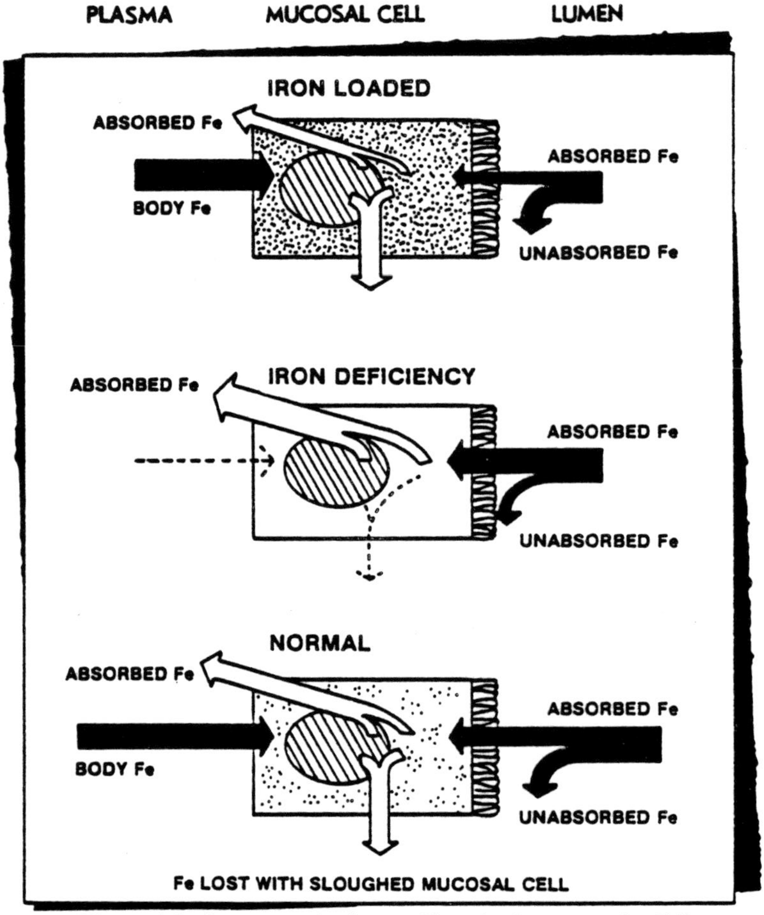

Figure 5-9. Storage and release of iron in the mucosal cell.[96]

cells and enters the plasma. Perhaps that is part of the motivation to move towards the basement membrane but osmosis is more probable, particularly because of active transport of small molecules across the membrane.[97] Water influx follows passively the actively transported molecules (ions, sugars, etc.) and the ferritin may be caught in this movement.

The concentration of iron in the plasma is less than the concentration in the mucosal cell. The concentration of iron on the exterior side of the brush border is also at a fairly high level. Consequently, with the cellular influx of water molecules from the plasma, there is a movement of ferritin towards the basement membrane to balance out the iron concentration.

One might even conceive this osmotic pressure as part of the regulatory mechanism for the entrance of iron into the blood. The greater the blood deficiency of the mineral, the stronger the pressure and the greater the pull and movement of mucosal cell absorbed iron towards the blood. When there is an adequate amount of plasma iron, there is little "pull" applied to the ferritin and the cell simply stores iron. This also occurs when its concentration is great enough to induce pressure towards the basement membrane or in cases of luminal deficiency through the brush border back into the lumen. In the course of storage in the terminal cell on the villus, if the cell dies and is sloughed off, the iron is lost. The osmotic pressure theory is quite speculative, but it does support the concepts of Figure 5-9.[96]

In summary, due to some as yet undefined mechanism, the iron, still chelated to the apoferritin molecule, is moved from the terminal web to the basement membrane. It can now be transported across the membrane into the plasma. Hendrix warns, "the mechanism that transports the solutes across the brush border membrane may be different from that effecting transport across the basal and lateral cell membranes".[87] Nevertheless, the same requirements for the entrance of the mineral into the cell apparently exist for the exit of the mineral except in reverse.

Moore and Dubach suggest that the rate of oxidation of the iron is greater near the mucosal cell surfaces.[97] This allows the ferrous ion to be changed to the trivalent state, released by the carrier protein, and rechelated with the apoferritin. Conversely at the other end of the cell, the oxidation potential would favor reducing the ferric iron in the ferritin molecule to the ferrous state. This change in ionic state could induce the release of the iron from the apoferritin molecule. At that point, the ferrous iron is ready to cross the basal cell membrane and enter into the plasma.[97]

It is assumed that the divalent iron atom is rechelated with a protein molecule in the basement membrane. If this is true, then the iron from the mucosal cell probably requires the flow of sodium and potassium in and out of the cell to be activated similar to the sodium-potassium pump which would move the membrane-bound iron to the cytoplasm. Again, there is an oxidation reduction reaction and the iron is oxidized to the ferric state. In this trivalent state, the iron could be chelated with either transferrin or siderophoron. Each molecule of that protein is able to chelate two atoms of iron by three phenolic groups. From there it is transported throughout the body.[98] Figure 5-10 summarizes the absorption of iron by the mucosal cell.[99]

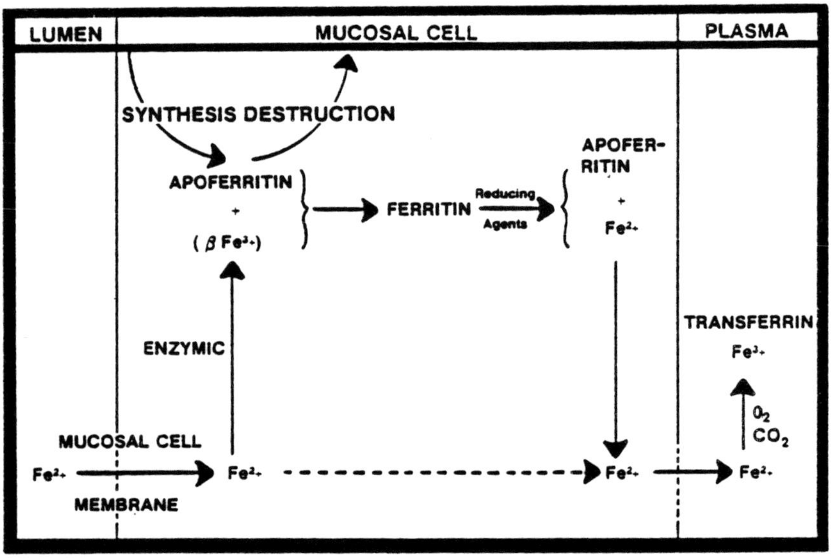

Figure 5-10. The movement of iron from the gut to the plasma.[99]

Like the metals previously discussed, mucosal uptake of ionic copper involves a two-step process.[100] This process involves the binding and the successive release of copper from macromolecular proteins found on the mucosal cell surface[101] through an enzymatic process.[102] The first step involves transporting copper from the lumen into the mucosal cell. The second step is concerned with moving the rebound and stored copper in the mucosal cell to the plasma.[103]

By a mechanism similar to that theorized by Evans for zinc absorption,[43] it has been proposed by some that copper is first bound to protein molecules in the lumen.[102,104] These proteins appear to be similar to the zinc-binding ligands, and it has therefore been postulated that copper follows the same absorption pathway as zinc.[105,106] In spite of the fact that copper is bound by these molecules, there are no data that conclusively demonstrate they are involved in intestinal absorption of copper any more than they are with zinc as was noted earlier.

There are data, however, that demonstrate that copper is chelated by ligands on the outer surface of the intestinal cell membrane.[38] It appears to be bound to either the carboxyl or the phosphoryl sites of the ligand. A sulfhydryl site may also be involved. A molar ratio of 1:1 (metal to ligand) is usually formed.[107] This appears to be similar to the integral proteins previously discussed.

Copper chelated to ligands on the mucosal cell surface would then be ready to be transported into the cytoplasm. This is accomplished by active transport.[108,109] What induces these ligands to give up the copper to protein molecules within the cell is not clear. Data presented by Rothstein and Meier[110] suggest the cellular proteins may have a greater affinity for the metal than the membrane surface ligands. Furthermore, there are differences in the behavior of these ligands to change in pH,[107] and different amino acids will form higher or lower stability constants with the same ion depending upon the pH of the surrounding medium.[111] There are measurable differences in the external and internal pH environments of the mucosal cell.[94]

Evans and LeBlanc[112] have isolated a copper-binding protein in the intestine. In the duodenum its molecular weight is about 10,000[105] although Riordan and Gower[113] identified three distinct forms of copper-binding proteins in liver cells having molecular weights of 7,00, 9,000 and 12,000. Their composition was similar to mucosal cell proteins.[112,113] It has been reported that copper is bound by metallothionein in the cell, but these copper-binding proteins had much less cysteine (7.5% to 6.8%) compared to the cysteine content of metallothionein.[114] With this exception, they are similar to metallothionein. Some of the properties of these copper-binding proteins are seen in Table 5-2.[114]

Table 5-2[114]

SOME PROPERTIES OF COPPER-BINDING PROTEINS

Molecular	Cu (g atoms/mole)	Cysteine (%)	Cysteine/Cu	Aromatic Amino Acid	Source
8,000	6	14.6	3.5	tyr, phe, his	Rat liver#
9,000‡	2.5	7.5	2.7	tyr, phe, his	Rat liver#
§	5.4	5.0	—	no	Rat liver#
§	0.8	20.4	—	no	Neonatal rat liver
§	§	2.6	—	tyr, phe, his	Rat intestine#
12,000	4.71¶	32	4.8	no	Pig liver# Fraction III-Ia
11,000	1.4	6.8	3.2	tyr, phe, his	Rat liver#
10,000	0.7	1.7	2.3	his	Yeast
10,000	1.8	5.1	2.7	no	Yeast
10,000	8.4	20.0	2.2	his	Yeast
10,000	10.0	24.3	2.2	his	Yeast
6,000	2.4	25	7.5	no	Human fetal liver
10,000	0.16*	4.6	46	tyr, phe, his	Normal human liver
10,000	0.32-4.3†	4.3	13.4-1	tyr, phe, his	Wilson's disease patient's liver

*Calculated based on a copper content of 0.1%

†Calculated based on a copper content of 0.2-2.7%.

‡Three proteins of 7,000, 9,000 and 12,000 were reported, but the average molecular weight (9,000) is given here and used for calculation of copper content and cysteine: Cu ratio.

§Not reported.

¶Calculated based on Cu/Zn ratio and cysteine/metal ratio.

#Animals injected with copper before protein isolation.

From these data it appears that uptake of ionic copper by mucosal cells involves chelation of the ion in the lumen by low molecular weight ligands located on the surface of the cell membranes, probably the integral proteins. The molar ratio of ligand to metal is about 1:1[107] Through an energy dependent process of active transport,[109] the copper is moved from the luminal side of the cells into their interiors. There it is released and rechelated with carrier proteins whose molecular weights approximate 10,000. They are rich in sulfhydryl groups and bind the copper through the formation of mercaptide bonds.[102] They are capable of binding a higher molar ratio of copper[114] and serve to store the metal in the mucosal cell as well as transport it to the plasma side.[102] This transport of copper through the mucosal cell to the plasma appears to involve a special mechanism or mechanisms that are dependent upon metabolic energy.[101]

Manganese ions are absorbed from the small intestine by a two-step mechanism similar to that of copper ions. The initial phase involves uptake from the lumen. This is followed by a transfer of the manganese across the mucosal cells to the plasma.[115] The manganese competes with both iron and cobalt ions for binding sites in both steps.[116,117] The iron-binding sites in the mucosa do not react with manganese to the same degree of specificty, however.[118] Manganese also competes with zinc ions for binding sites on the intestinal cell surface.[119] This suggests a common pathway.

The manganese ion is first bound to the surface of the mucosal cell.[91] As suggested by Rothstein,[107] the bonding occurs at the carboxyl, the sulfhydryl, and/or the phosphoryl sites and forms a chelate with a ligand to metal molar ratio of 1:1.

Once the manganese has entered the mucosal cell by active transport,[120] it is released and rechelated by protein molecules found within the cell.[115] As was discussed in relation to the membrane, transport of copper and its subsequent release and uptake by cellular proteins, this occurrence is probably induced by both changes in pH[94,107] and the introduction of ligands with higher stability constants.[110] "The behavior of any one metal is dependent upon its immediate environment in which the presence of other metal ions as well as other molecular species play an important part."[121]

It appears that the proteins in the mucosal cells can affect the second step in the manganese absorption process by first storing the manganese before transferring it to the plasma on the serosal side.[31] Nevertheless, the transfer to the serosa *in vitro* was much more rapid for the manganese compared to zinc. This would suggest that manganese and zinc share a common transport mechanism, but metal transport studies by the perfusion method indicate that the transport of manganese has no inhibitory effect on zinc transport.

Thus, Sahagian, *et al.* concluded, "metal uptakes are affected by the presence of other metal ions. Some of the binding sites are probably common to more than one metal with the binding affinity varying from metal to metal. Firm binding of a metal by the tissue may lead to considerable uptake but little transport and vice versa. The transport of a given metal, or its accumulation in the serosal solution, is largely dependent upon the degree of affinity of the metal for binding sites available on the tissue, either on the outer or inner surfaces, and the total capacity of the tissue for binding or retaining a given metal. Therefore, any factor which activates tissue-binding sites or alters the binding characteristics of the metallic ion by virtue of chelation, competition, or some other means, results in lower tissue retention of that metal and hence enhances transport across that tissue and accumulation in the solution bathing the opposite side. Moreover, the possibility also exists that the inhibitor or accelerator factor by combining simultaneously with the metallic ion and the tissue site may bring about an increased tissue retention of a particular metallic species".[122]

In summary, it can be seen that the intestinal absorption mechanism of divalent metal ions is very similar: Calcium is bound by proteins[123] in the cell membrane in a very specific geometric pattern[124] which is presumably a ring structure. Uptake is accomplished by an active transport system.[125] Zinc is tenaciously bound to a protein molecule that has a molecular weight of about 800[126] which is smaller than serum albumin.[127] This is not a simple amino acid but a peptide.[128] Absorption is by an active process requiring energy.[129] It is believed that magnesium shares a common transport mechanism with calcium[130] and if so, would require an amino acid ligand for active transport across the membrane.[131] The binding of magnesium requires the formation of a coordinate bond with the membrane protein which is necessary for chelation on the cell membrane.[132] Iron is bound to protein in the mucosal cell.[133] This protein ligand has a low molecular weight,[134,135] and through the process of chelation prevents or retards the inevitable precipitation of the ingested iron at the alkaline pH of the small intestine.[136] Like magnesium, copper ions are bound to protein molecules on the mucosal cell surface.[137,139] These proteins are of a low molecular weight.[138] After binding, the absorption of the copper chelate is energy dependent.[137] And finally, manganese is also bound to the intestinal membrane.[119] It is believed to be absorbed by a mechanism that is similar to the absorption of iron.[120] This requires not only the protein ligand, but also active transport across the membrane.[131]

The similarity in mucosal cell membrane transport mechanisms is summarized in Table 5-3.

Table 5-3 strongly suggests that the mineral ions under consideration all behave in a similar manner during intestinal absorption. Their absorption behavior and requirements are so similar that similar pathways are indicated.

Table 5-3.

SIMILARITIES IN MUCOSAL MEMBRANE TRANSPORT OF IONS

	Ca	Mg	Fe	Zn	Cu	Mn
Absorption in acid pH of upper intestine	X	X	X	X	X	X
Solubility/ionization essential for absorption	X	X	X	X	X	X
Ion attached to exterior cell membrane	X	X	X	X	X	X
Membrane attachment requires amino acid/peptide molecule	X	X	X	X	X	X
Ion binding to peripheral protein requires chelation	X	X	X	X	X	X
Membrane transport requires energy	X	X	X	X	X	X
Mineral released intra-cellularly and rechelated	X	X	X	X	X	X

REFERENCES

1. Evans, G., "Zinc Absorption and Transport," in Prasad, A., ed., *Trace Elements in Human Health and Disease* (New York: Academic Press), V1, 181, 1976.
2. Memmler, R. and Wood, D., *Structure and Function of the Human Body* (Philadelphia: J. B. Lippincott Co.), 142, 1977.
3. Code, C. F., ed., *Handbook of Physiology* (Baltimore: William & Wilkins Co.), V3, 1125, 1968.
4. Waldron-Edward, D., "Effects of pH and Counter-Ion on Absorption of Metal Ions," in Skoryna, S. and Waldron-Edward, D., eds., *Intestinal Absorption of Metal Ions, Trace Elements and Radionuclides* (Oxford: Pergamon Press), 381, 1971.
5. Wasserman, R., and Taylor, A., "Some Aspects of the Intestinal Absorption of Calcium with Special Reference to Vitamin D," in Comar, C. and Bronner, F., eds., *Mineral Metabolism, and Advanced Treatise* (New York: Academic Press), V3, 345, 1969.
6. Cramer, D., "Sites of Calcium Absorption and the Calcium Concentration of Gut Contents in the Dog," Can. J. Physiol. Pharmacol., 43:75, 1965.

7. Pauling, L., *College Chemistry* (San Francisco: W. H. Freeman and Co.), 444, 1955.
8. Ebel, J., *et al.*, "Vitamin D-Induced Calcium-Binding Protein of Intestinal Mucosa," Am. J. Clin. Nutr., 22:431, 1969.
9. Wilson, E., *Principles of Nutrition* (New York: John Wiley & Sons), 262, 1979.
10. Guthrie, H., *Introductory Nutrition* (St. Louis: C. V. Mosby Co.), 116, 1975.
11. Ibid., 174.
12. Cartwright, G., *et al.*, "Studies on Copper Metabolism," J. Clin. Invest., 33:1487, 1954.
13. Scheinberg, H. and Sternlieb, I., "Copper Toxicity and Wilson's Disease," in Prasad, A., ed., *Trace Elements in Human Health and Disease* (New York: Academic Press), V1, 416, 1976.
14. Lee, G., "Role of Copper in Iron Metabolism and Heme Biosynthesis," in Prasad, A., ed., *Trace Elements in Human Health and Disease* (New York: Academic Press), V1, 373, 1976.
15. Owen, C., "Absorption and Excretion of Cu^{64}-Labelled Copper by the Rat," Am. J. Physiol., 207:1203, 1964.
16. Tompsett, S., "Factors Influencing the Absorption of Iron and Copper from the Alimentary Tract," Biochem. J., 34:961, 1940.
17. Van Campen, C., and Mitchell, E., "Absorption of Cu^{64}, Zn^{65}, Mn^{99} and Fe^{59} from Ligated Segments of the Rat Gastrointestinal Tract," J. Nutr., 86:120, 1965.
18. Classen, H., *et al.*, "Experimental Studies on the Intestinal Absorption of Magnesium and its Protective Effects Against Cardiac Hypertrophy and Nonocclusive Necroses," Paper given at Second International Symposium of Magnesium, Montreal, Quebec, May 30, 1976.
19. Chutkow, J., "Sites of Magnesium Absorption and Excretion in the Intestinal Tract of the Rat," J. Lab. Clin. Med., 63:71, 1964.
20. Briggs, G. and Calloway, D., *Nutrition and Physical Fitness* (Philadelphia: W. B. Saunders Co.), 265, 1979.
21. Forth, W. and Rummel, W., "Absorption of Iron and Chemically Related Metals *In Vitro* and *In Vivo*: Specificity of the Iron Binding System in the Mucosa of the Jejunum," in Skoryna, S. and Waldron-Edward, D., *Intestinal Absorption of Metal Ions, Trace Elements and Radionuclides* (Oxford: Pergamon Press), 181, 1971.
22. Brown, E. and Justus, B., "*In Vitro* Absorption of Radioiron by Everted Poaches of Rat Intestine," Am. J. Physiol., 194:319, 1958.
23. Sharpe, L., *et al.*, "The Effect of Phytate and Other Food Factors on Iron Absorption," J. Nutr., 41:433, 1950.
24. Moore, C. and Dubach, R., "Metabolism and Requirements of Iron in the Human," J. Am. Med. Assoc. 162:197, 1956.
25. Oelshlegel, F. and Brewer, G., "Absorption of Pharmacological Doses of Zinc," in Brewer, G. and Prasad, A., eds., *Zinc Metabolism* (New York: Alan R. Less), 310, 1977.
26. Reinhold, J. *et al.*, "Binding of Zinc to Fiber and Other Solids of Wholemeal Bread," in Prasad, A., ed., *Trace Elements in Human Health and Disease* (New

Negative charge on wall
of intestine for Max

To DeWayne

Time ____________ Date ____________

While You Were Out

M CJ Lane

Of is holding on 7600

Phone No. ____________

☐ Telephoned ☐ Please Call

☐ Returned Your Call ☐ Will Call Again

☐ Left The Following Message:

Call Received By ____________

York: Academic Press), V1, 169, 1976.

27. Van Campen, C., and Mitchell, E., "Absorption of Cu^{64}, Zn^{65}, Mn^{49} and Fe^{59} from Ligated Segments of the Rat Gastrointestinal Tract," J. Nutr., 86:120, 1965.
28. Underwood, E., *Trace Elements in Human and Animal Nutrition* (New York: Academic Press), 205-6, 1977.
29. Florkin, M. and Statz, E., eds., *Comprehensive Biochemistry* (Amsterdam: Elsevier Publishing Co.), V21, 223.
30. Thompson, A., *et al.*, J. Lab. Clin. Med., 78:642, 1971.
31. Sahagian, B., *et al.*, "Effects of Inhibitors and Accelerators on Intestinal Absorption of Divalent Trace Elements," in Skoryna, S. and Waldron-Edward, D., eds., *Intestinal Absorption of Metal Ions, Trace Elements and Radionuclides* (Oxford: Pergamon Press), 326-328, 1971.
32. Scatchard, G., *et al.*, "Physical Chemistry of Protein Solutions. X. The Binding of Small Anions by Serum Albumen," J. Am. Chem. Soc., 81:6104, 1959.
33. Saroff, H., "A Theory for the Binding of Chloride Ions to Serum Album Based on a Hydrogen Bonded Model," J. Phys. Chem., 61:1364,, 1957.
34. Scatchard, G. and Black, E., "The Effect of Salts on the Isoionic and Isoelectric Points of Proteins," J. Phys. & Colloid. Chem., 53:88, 1949.
35. Rose, R., "Absorptive Function of the Gallbladder," in Johnson, L., *et al.*, ed., *Physiology of the Gastrointestinal Tract* (New York: Raven Press), V2, 1022, 1981.
36. Case, R., *et al.*, "The Origin and Secretion of Pancreatic Juice Bicarbonate," J. Physiol., 210:1, 1970.
37. Malagelada, J., "Gastric, Pancreatic, and Biliary Responses to a Meal," in Johnson, L., *et al.*, eds., *Physiology of the Gastrointestinal Tract* (New York: Raven Press), V2, 911, 1981.
38. Donsbach, K., "Digestive System," (A filmstrip produced by Donsbach University, Huntington Beach, CA), 1980.
39. Schachter, D., *et al.*, "Active Transport of Calcium by the Small Intestine of the Rat," Am. J. Physiol., 198:263, 1960.
40. Wilson, E., *Principles of Nutrition* (New York: John Wiley & Sons), 262, 1979.
41. Ibid, 273.
42. Forth, op. cit., 174.
43. Evans, G., "Zinc Absorption and Transport," in Prasad, A., ed., *Trace Elements in Human Health and Disease* (New York: Academic Press), V1, 183-5, 1976.
44. Thompson, A., *et al.*, "Interrelation of Intestinal Transport System for Manganese and Iron," J. Lab. Clin. Med., 78:642, 1971.
45. Tomsett, S., "Factors Influencing the Absorption of Iron and Copper from the Alimentary Tract," Biochem. J., 34:961, 1940.
46. Huebers, H. and Rummel, W., "Iron Binding Proteins: Mediators in Iron Absorption," in Kramer, M. and Lauterbach, F., eds., *Intestinal Permeation* (Amsterdam: Excerpta Medica), 378, 1977.
47. Pauling, L., *The Nature of the Chemical Bond* (Ithaca: Cornell University Press),

61, 1960.
48. Ibid., 6.
49. Mellor, D., "Historical Background and Fundamental Concepts," in Dwyer, F. and Mellor, D., eds., *Chelating Agents and Metal Chelates* (New York: Academic Press), 1, 1964.
50. Wasserman, R., *et al.*, ed., *Calcium Binding Proteins and Calcium Function* (Amsterdam: Elsevier—North Holland), 292, 1977.
51. Zile, M., *et al.*, "Localization of 1,25-Dehydroxyvitamin D_3 in Intestinal Nuclei *in Vivo*," Arch. Biochem. Biophys., 186:15, 1978.
52. DeLuca, H., "Vitamin D Metabolism and Function," in Kharasch, N., ed., *Trace Metals in Health and Disease* (New York: Raven Press), 209, 1979.
53. Wasserman, R., *et al.*, "Vitamin D-Dependent Calcium-Binding Protein: Purification and Some Properties," J. Biol. Chem., 243:3978, 1968.
54. Ebel, J., *et al.*, "Vitamin Mucosa," Am. J. Clin. Nutr., 22:431, 1969.
55. Hastings, A., *et al.*, "The Ionization of Calcium, Magnesium and Strontium Citrates," J. Biol. Chem., 107:351, 1934.
56. Weir, E. and Hastings, A., "The Ionization Constants of Calcium Proteinate Determined by the Solubility of Calcium Carbonate," J. Biol. Chem., 114:397, 1936.
57. Briggs, F. and Fleishman, M., "Calcium Binding by Particle-Free Supernatants of Homogenates of Skeletal Muscle," J. Gen. Physiol., 49:131, 1965.
58. Taylor, A. and Wasserman, R., "Vitamin D_3-Induced Calcium-Binding Protein: Partial Purification, Electrophoric Visualization and Tissue Distribution," Arch., Biochem. Biophys., 119:536, 1967.
59. Wasserman, R., "Studies on Vitamin D_3 and the Intestinal Absorption and Other Ions in the Rachitic Chick," J. Nutr., 77:69, 1962.
60. Zull, J., *et al.*, "On the Relationship Between Vitamin D Action and Actinomycin-Sensitive Processes," Proc. Nat. Acad. Sci., 55:177, 1966.
61. Taylor, A., *Vitamin D_3 Induced Calcium-Binding Protein: Immunochemical and Florescent Antibody Localization Studies*, Ph.D. Thesis, Cornell University, Ithaca, New York, 1969.
62. Wasserman and Taylor, op. cit., 377.
63. Ibid., 378.
64. Hsu, J., "Biochemistry and Metabolism of Zinc," in Karcioglu, Z. and Sarper, R., eds., *Zinc and Copper in Medicine* (Springfield: Charles C Thomas), 75, 1980.
65. Hahn, C., *et al.*, "Structure of a Zinc Ligand Isolated from Duodenum," Fed. Proc., 35:361, 1976.
66. Evans, G., *et al.*, "A Proposed Mechanism for Zinc Absorption in the Rat," Am. J. Physiol., 228:501, 1975.
67. Song, M. and Adham, N., "A Possible Role for a Prostagland-Like Substance in Zinc Absorption," Fed. Proc., 35:1667, 1976.
68. Song, M. and Adham, N., "Prostaglandin Regulation of Zinc Absorption in Rats," Fed. Proc., 36:1138, 1977.
69. Schricker, B., *Studies of an Endogenous Zinc Chelator of Rat Intestine*, Ph.D. Thesis, University of Illinois, 1978.

70. Hahn, C. and Evans, G., "Identification of a Low Molecular Weight ^{65}Zn Complex in Rat Intestine," Proc. Soc. Exp. Biol. Med., 144:793, 1973.
71. Van Campen, D. and Kowalski, T., "Studies on Zinc Absorption: ^{65}Zn Binding by Homogenates of Rat Intestinal Mucosa," Proc. Soc. Exp. Med., 136:294, 1971.
72. Kagi, J. and Vallee, V., "Metallothionein: A Cadmium and Zinc Containing Protein from Renal Cortex," J. Biol. Chem., 235:3460, 1960.
73. Kienholz, E., *et al.*, "Influence of Dietary Zinc, Calcium and Vitamin D for Hens on Zinc Content of Tissues and Eggs and on Bone Composition," Poultry Sci., 43:667, 1964.
74. Whiting, F. and Bezeau, L., "The Calcium, Phosphorus and Zinc Balance in Pigs as Influenced by the Weight of Pig and the Level of Calcium, Zinc and Vitamin D in the Ration," Can. J. Ani. Sci., 38:109, 1958.
75. Richards, M. and Cousins, R., "Influence of Parenteral Zinc and Actinomycin D on Tissue Zinc Uptake and the Synthesis of a Zinc-Binding Protein," Bioinog. Chem., 4:215, 1975.
76. Richards, M. and Cousins, R., "Mammalian Zinc Homostosis: Requirement for RNA and Metallothionein Synthesis," Biochem. Biophys. Res. Comm., 64:1215, 1975.
77. Squibb, K., *et al.*, "Control of Zinc-Thionein Synthesis in Rat Liver," J. Biochem., 164:223, 1977.
78. Petith, M. and Schedl, H., "Magnesium and Calcium Transport in Rat Large Intestine *In Vivo* Adaptation to Calcium and Magnesium Deficiency," in Cantin, M. and Seelig, M., eds., *Magnesium in Health and Disease* (New York: SP Medical & Scientific Books), 153, 1980.
79. Heaton, F., *et al.*, "Observations in the Relationships Between Calcium and Magnesium Metabolism in Man," Clin. Sci., 27:31, 1964.
80. Brickman, A., *et al.*, "Action of $1,25(OH)_2$ and 1(OH)-Vitamin D_3 on Magnesium Metabolism in Man," Clin. Res., 23:315, 1975.
81. MacIntyre, I., "Discussion on Magnesium Metabolism in Man and Animals," Proc. Royal Soc. Med., 53:1037, 1960.
82. Hendrix, J., *et al.*, "Competition Between Calcium, Strontium, and Magnesium for Absorption in the Isolated Rat Intestine," Clin. Chem., 9:734, 1963.
83. Cohen, H., "Purgative Action of Magnesium Salts," Quart. J. Med., 19:249, 1926.
84. Tansy, M., "Intestinal Absorption of Magnesium," in Skoryna, S. and Waldron-Edward, D., eds., *Intestinal Absorption of Metal Ions, Trace Elements and Radionuclides* (Oxford: Pergamon Press), 196-7, 1971.
85. Ibid., 199.
86. Freeman, J., and Geer, J., "Intestinal Fat and Iron Transport, Goblet Cell Mucus Secretion, and Cellular Changes in Protein Deficiency Observed with the Electron Microscope," Am. J. Digestive Disease 10:1116, 1965.
87. Hendrix, T., "The Absorptive Function of the Alimentary Canal," in Mountcastle, V., ed., *Medical Physiology* (St. Louis: C. V. Mosby), V2,, 1145-7, 1974.

88. Manis, J. and Schackter, D., "Active Transport of Iron by Intestine," Amer. J. Physiol. 203:73, 1962.
89. Granick, S., "Iron Transport Mechanism in the Intestinal Tract," Ross Pediatric Res. Conf., Oct. 20-21, 1965.
90. Thompson, A., *et al.*, "Interrelation of Intestinal Transport System for Manganese and Iron," J. Lab. Clin. Med., 78:642, 1971.
91. Underwood, op. cit., 62.
92. Ogata, E., *et al.*, "Basis of Action of Hormones on Calcium Absorption," in Skoryna, S. and Waldron-Edward, D., eds., *Intestinal Absorption of Metal Ions, Trace Elements and Radionuclides* (Oxford: Pergamon Press), 359, 1971.
93. Jensen, N., "Biological Assimilation of Metals," U.S. Patent No. 4,167,564, Washington, D.C., Sept. 11, 1979.
94. Crim, M. and Munro, H., "Protein", in Hegsted, D., *et al.*, ed., *Present Knowledge in Nutrition* (New York: The Nutrition Foundation), 44, 1976.
95. Underwood, op. cit., 21.
96. Hendrix, op. cit., 1171.
97. Moore, C. and Dubach, R., "Iron," in Comar, C. and Bronner, F., eds., *Mineral Metabolism: An Advanced Treatise* (New York: Academic Press), V2, Pt. B, 305, 1962.
98. Davies, B., *et al.*, "The Stability Constants of the Iron Transferrin Complex," Biochem. Biophys. Res. Commun., 8:56, 1962.
99. Granick, S., "Structure and Physiological Functions of Ferritin," Physiol. Rev., 31:497, 1951.
100. Gitlin, D., *et al.*, "Absorption and Excretion of Copper in Mice," Nature, 188:150, 1960.
101. Van Campen, D., "Absorption of Copper from the Gastrointestinal Tract," in Skoryna, S. and Waldron-Edward, D., eds., *Intestinal Absorption of Metal Ions, Trace Elements and Radionuclides* (Oxford: Pergamon Press), 213, 1971.
102. Mason, K., "A Conspectus of Research on Copper Metabolism and Requirements of Man," J. Nutr., 109:1990, 1979.
103. Hunt, D., Nature, 249:852, 1974.
104. Friedman, M., ed., *Protein-Metal Interactions* (New York: Plenum), 1974.
105. Starcher, B., "Studies on the Mechanism of Copper Absorption in the Chick," J. Nutr., 97:321, 1969.
106. Van Campen, D., "Copper Interference with the Intestinal Absorption of ^{65}Zn by Rats," J. Nutr., 97:104, 1969.
107. Rothstein, A., "A Reappraisal of the Action of Uranyl Ion on Cell Membranes," in Maniloff, J., *et al.*, ed., *Effects of Metals on Cells, Subcellular Elements, and Macromolecules* (Springfield: Charles C Thomas), 365, 1970.
108. O'Dell, B. and Campbell, B., "Trace Elements: Metabolism and Metabolic Function," in Florkin, M., and Stotz, E., eds., *Comprehensive Biochemistry* (Amsterdam: Elsevier Publishing Co.), V21, 192, 1971.
109. Kirchgessner, M. and Grassman, E., "The Dynamics of Copper Absorption," in Mills, C., ed., *Trace Element Metabolism in Animals* (Edinburgh: Livingstone), 277, 1970.
110. Rothstein, A. and Meier, R., "The Relationship of the Cell Surface to

Metabolism," J. Cell. Comp. Physiol., 38:245, 1951.
111. Bjerrum, J., *Metal Ammine Formation in Aqueous Solution* (Copenhagen: Haase), 1941.
112. Evans, G. and LeBlanc, F., "Copper Binding Protein in Rat Intestine: Amino Acids Composition and Function," Nutr. Rept. Int., 14:281, 1976.
113. Riordan, J. and Gower, I., "Purification of Low Molecular Weight Copper Proteins from Copper Loaded Liver," Biochem. Biophys. Res. Commun., 66:678, 1975.
114. Hsieh, S. and Hsu, J., "Biochemistry and Metabolism of Copper," in Karcioglu, Z. and Sarper, R., eds., *Zinc and Copper in Medicine* (Springfield: Charles C Thomas), 109-10, 1980.
115. Underwood, op. cit., 175.
116. Thompson, A., *et al.*, "Interrelation of Intestinal Transport System for Manganese and Iron," J. Lab. Clin. Med., 78:642, 1971.
117. Thompson, A. and Valberg, L., "Intestinal Uptake of Iron, Cobalt and Manganese in the Iron-Deficient Rat," Am. J. Physiol., 223:1327, 1971.
118. Forth, W., "Absorption of Iron and Chemically Related Metals *In Vitro* and *In Vivo*; the Specificity of an Iron Binding System in the Intestinal Mucosa of the Rat," in Mills, C., ed., *Trace Element Metabolism in Animals* (Edinburgh: E & S Livingstone), 307, 1970.
119. Sahagian, op. cit., 330-1.
120. Guthrie, op. cit., 174.
121. Sahagian, op. cit., 323.
122. Ibid., 333-34.
123. Tobias, J., *et al.*, "The Excitable System in the Cell Surface," Circulation 26:1145, 1962.
124. DeGrazia, J., "The Intestinal Absorption of Calcium," in Skoryna, S. and Waldron-Edward, D., eds., *Intestinal Absorption of Metal Ions, Trace Elements and Radionuclides* (Oxford: Pergamon Press), 152, 1971.
125. Rasmussen, H., "Mitochondrial Ion Transport. Mechanism and Physiological Significance," Fed. Proc., 25:903, 1966.
126. Sing, M. and Adham, N., "Role of Prostaglandin E2 in Zinc Absorption in the Rat," Am. J. Physiol., 234:599, 1978.
127. Suso, F. and Edwards, H., "Binding Capacity of Intestinal Mucosa and Blood Plasma for Zinc," Proc. Soc. Exp. Biol. Med., 137:306, 1971.
128. Hahn, C. and Evans, G., "Identification of a Low Molecular Weight ^{65}Zn Complex in Rat Intestine," Proc. Soc. Exp. Biol. Med., 144:793, 1973.
129. Solomons, N., "Zinc and Copper in the Gastrointestinal System," in Karcioglu, Z. and Sarper, R., eds., *Zinc and Copper in Medicine* (Springfield: Charles C Thomas), 277, 1980.
130. Alcock, N. and MacIntyre, I., "Interrelationships of Calcium and Magnesium Absorption," Clin. Sci., 22:185, 1962.
131. Scheiner, E., *et al.*, "Malabsorption Following Massive Intestinal Resection," Amer. J. Clin. Nutr., 17:64, 1965.
132. Sahagian, op. cit., 331.
133. Jandl, J. and Katz, J., "The Plasma-to-Cell Cycle of Transferrin," J. Clin.

Invest., 42:314, 1963.

134. Blanc, B., "Etude Isotopique des Proteines Implequees dans la Resorption et dans le Metabolisme du Fer," Vortrag, gehalten auf der 3 Jahrestagung des Ges. f. Nuklearmedizin in Lausanne, 21-23 Oct. 1965.
135. Vannoti, R., and Hoffman, W., *Radioisotope in der Gastroenterologie* (Munich: F. K. Schaattauer), 113, 1967.
136. Beutler, E., "Iron," in Goodhart, R. and Shils, M., eds., *Modern Nutritionin Health and Disease* (Philadelphia: Lea and Febiger), 327, 1980.
137. Crampton, R., *et al.*, "Observations on the Mechanism of Absorption of Copper by the Small Intestine," J. Physiol., 178:111, 1965.
138. Evans, G., *et al.*, "A Proposed Mechanism for Zinc Absorption in the Rat," Am. J. Physiol., 228:501, 1975.
139. Persach, P. and Blumberg, W., eds., *The Biochemistry of Copper* (New York: Academic Press), 183, 1966.

CHAPTER 6

ABSORPTION OF METAL IONS BY ACTIVE TRANSPORT

IN Chapter 5 it was shown that for an inorganic mineral salt to be absorbed into the mucosal cell it must first be ionized in the lumen of the gut. This step is essential because the mineral must be in the ionic state in order to be chelated with protein molecules in the membrane of the mucosal cell. These integral proteins, which are products of cellular metabolism,[1] then act as transport mechanisms to carry the cation from the luminal side of the membrane into the cytoplasm. The movement of the metal ion into the cell is accomplished through an active transport process and illustrated in Figure 6-1.[1]

In Chapter 3 a model of a plastic dynamic membrane which was proposed by Singer and Nicolson[3] was presented. As seen in Figure 6-2[4], this membrane consists of two layers of lipids. The hydrophilic ends are pointed outwards so fluids can be absorbed from the external side of the membrane. The hydrophobic ends of the lipids point inwards. This tends to prevent the backward flow of fluids out of the cell.

The peripheral proteins and integral proteins are embedded in the lipid layer. The glycocalyx is attached to the protein molecules. Composed of glycoproteins and glycolipids, the protein-to-lipid ratio is approximately 1.7:1.[5] The glycocalyx is charged and protrudes from the surfaces of the peripheral and integral proteins.[4] It is important in the absorptive processes. For example, it selectively binds Ca^{-} and presumably other cations in the intestine.[6,7]

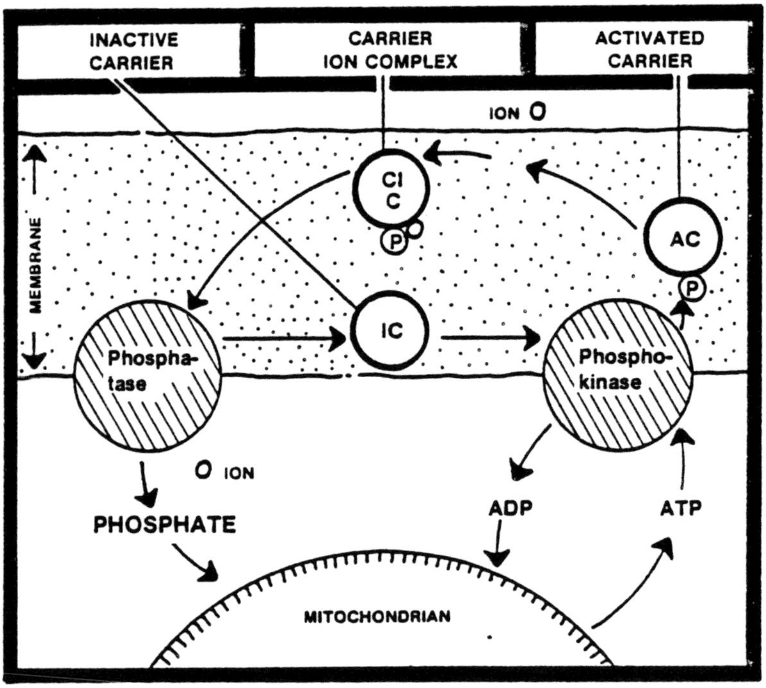

Figure 6-1. The movement of an ion into the cell.

"This structure allows the fluidity which is necessary for passage of molecules through the membrane. Membrane components are redistributed 'by translational diffusion' through the viscous two-dimensional solutions, thereby permitting new interactions among the altered components. Cell membrane fluidity, accompanied by alterations in the localization of specific membrane components, has been reported. . . . Alterations in shape, as well as the apparent 'fluidity' of plasma membranes, may be attributed to a major degree to the presence therein of an actinomysin-like system that resembles the contractile actinomysin component of smooth muscle."[8]

In order for an ion to move through this membrane it must be attached to a carrier protein found in the brush border of the membrane which carries it through the membranes.[25] This is because the charge on the cation prohibits its passage through the entry channels of the membrane which are also charged. The chelation of the ion by the carrier protein neutralizes the ionic charge. On a frequency distribution basis the highest number of intestinal cells with COOH

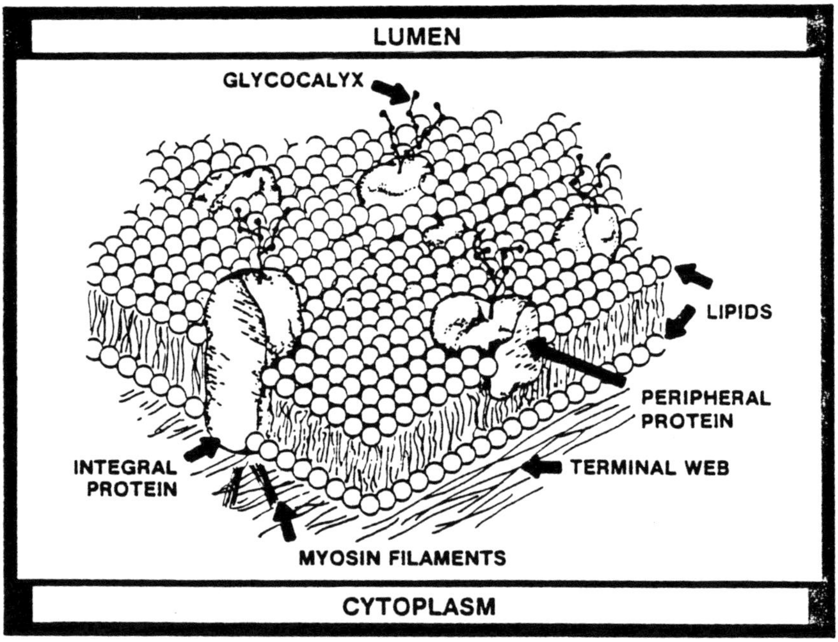

Figure 6-2. Model of a mucosal cell membrane.

terminals for chelation purposes is in the duodenum and the first part of the jejunum.[28] Once chelated, some ion movement may occur as a result of facilitated diffusion although most movement across the membrane is due to active transport; the two systems can operate simultaneously.

The carrier proteins are highly specific and have binding specifications that are different. Carrier proteins have been identified for calcium, sodium, and potassium. Each has a molecular weight of about 30,000 and a dissociation constant breakdown into subunits of 10^{-3} to 10^{-6}. Although their behaviors are similar, their amino acid compositions are different. These carriers are probably part of the glycocalyx and extend into the lumen. They participate in both facilitated diffusion along a concentration gradient and in active transport uphill against a concentration gradient. In the second case they function only when energy is provided.[12]

Active transport is defined as an "enzyme-catalyzed reaction occurring on one side of the membrane that is linked to the vectoral transport of a substance not chemically altered by the process".[2]

Metabolic energy (ATP) derived from oxidative phosphorylation in the mitochondria is provided to the carrier protein molecules which then move the ion across the membrane probably through a series of conformational changes.[10,11] (see Figure 3-12.)

Active transport allows the mucosal cell to absorb a particular ion at a rate determined solely by the needs of the cell regardless of the concentration of the ion in the lumen. Thus it has the advantage over facilitated diffusion of not requiring a favorable concentration of the ion in the lumen or in the cell. However, as stated earlier, when the concentration of a specific ion is high in the lumen, facilitated diffusion may move some of it into the cell at the same time it is being moved by active transport. On the other hand, when the external concentration is lower than the internal concentration, active transport is the sole purveyor of ion entry into the cell.[12] White, *et al.*,[8] believe this system of supplying energy for active transport is somehow correlated with cellular metabolic demands.

There is no well-defined model that demonstrates the movement of all ions across the membrane by active transport. A phosphate pump has been identified.[25] The sodium-potassium pump is the most widely studied. A calcium pump has also been identified and most widely examined in the sarcoplasmic reticulum, although Martonosi[18] has indicated that its operation appears to be identical in the surface membranes of animal cells. Although the sodium and calcium pumps are not identical, they are very similar. Thus, for the purpose of supplying a general overview of a possible biochemical pumping mechanism for cations, this discussion will borrow from both.

Structurally the calcium transport carrier appears to consist of an oligomer. It has two or more large subunits measuring approximately 100,000 daltons each. A small proteolipid measuring about 12,000 daltons and about 30 moles of phospholipid may be associated with each large subunit. There may be a 55,000 dalton glycoprotein associated with it as well. This glycoprotein may be necessary for the normal transport of the calcium.[15] It has a high-binding affinity for the calcium.[16] The glycoprotein appears to function as an anion in the chelation of the calcium on the external side of the membrane,[14] and in fact may be part of the glycocalyx found on the microvilli.

In reseach reported by Miyamoto and Kasai,[17] the binding capacity was determined to be 11.9 + .9 moles per mg of protein. This corresponds to about 1.5 Ca" cations bound per subunit.[14]

Physiologically the transport carrier probably traverses the membrane bilayer. It chelates the calcium cation on the luminal side of the mucosal cell. Then reacting with ATP furnished from the cytoplasmic side, a series of molecular configuration changes in the carrier probably occur and the cation is passed from molecule to molecule, thus traversing the membrane. (See Figure 6-5) The binding of the calcium to the glycoprotein appears to trigger the release of cytoplasmic ATP to move the cation.

Energy is derived from the hydrolysis of ATP by ATPase. The hydrolysis of ATP involves "A ($Ca^{2+} + Mg^{2+}$) dependent phosphorylation of the B-carboxyl of an aspartyl residue of the large subunit, followed by an Mg^{2+} -dependent hydrolysis of the phosphoryl enzyme. (Ca^{2+} transport is considered to result from a cycle of conformational transitions driven by the phosphorylation-dephosphorylation cycle as in the Na^{2+} system."[14] The ATPase is a single polypeptide with a molecular weight of between 120,000 and 130,000. Functional ATPase in the membrane is, however, associated with dimers or oligomers. "The ATP modulations could then result from ATP binding at a catalytic site on one enzyme (through conformational interactions) helping to drive rate-limiting steps at another catalytic site. Catalytic sites would then operate in a coordinated alternating manner."[19]

"The ($Na^+ + K^+$) ATPase [also] uses the hydrolysis of ATP as the source of energy for the active, coupled transport of Na^+ outward and K^+ inward across a wide variety of cell membranes. It requires the presence of Na^+, K^+, Mg^{2+}, and ATP, and is thought to operate via mechanism similar to the one shown in Figure 6-3 where E_1 and E_2 are different conformations of the phosphorylated enzyme, the former of higher energy and sensitive to ADP, which reverses the reaction, and the latter of lower energy and sensitive to K^+, which catalyzes its breakdown to E_2. It is still a matter of some controversy at exactly which step(s) the ion transport occurs, and whether the transports of Na^+ and K^+ are simultaneous or occur at different stages. However, there is strong evidence that phosphorylation of the enzyme is Na^+ dependent and dephosphorylation is K^+ dependent and the K^+ inhibits the conversion of E_2 to E_1. Thus, the kinetic evidence is consistent with the transports of Na^+ and K^+ occurring at separate steps."[14]

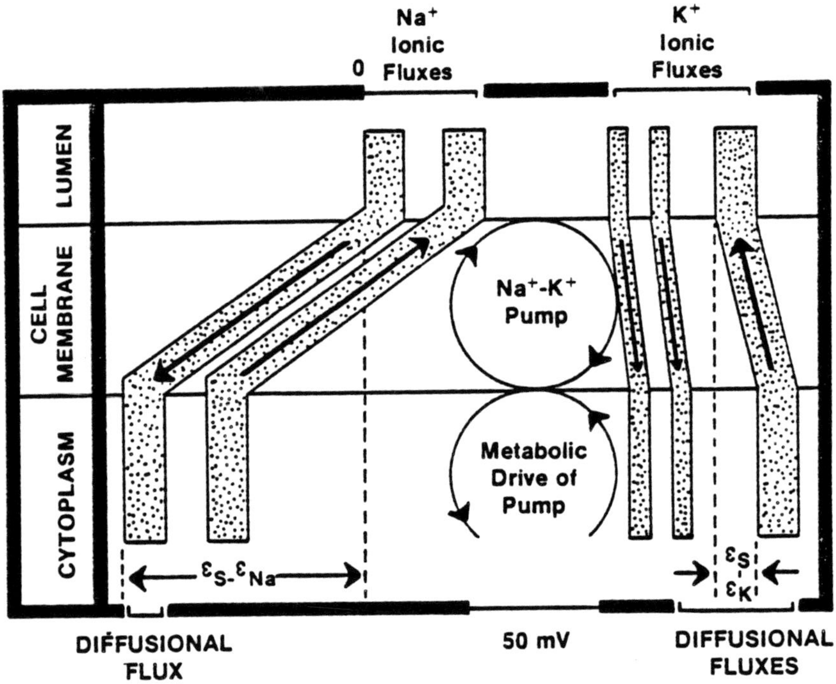

Figure 6-3. A sodium-potassium pump in the cell membrane.[14]

In essence then, the mechanism or "pump" that moves the various ions across the cell membrane can probably be regarded as an enzyme (ATPase). The ATP is the substrate, and the various cations mentioned above function as cofactors in the enzyme system. Glynn and Karlish[20] have reported that in experiments with intact cells the alkali-metal ions must be considered as either substrates or as products, depending upon which side of the membrane they are on. Thus, they believe that the movement of ions across the membrane is equivalent to the conversion of a substrate to a product. Two such models are seen in Figures 6-4[21] and 6-5[13].

In Figure 6-4[21] the movement of sodium and potassium is illustrated. That movement affects the hydrolysis of ATP by ATPase. In Figure 6-5[13] the sodium-potassium pump is not as detailed, but the movement of the ion across the membrane and the various enzymatic reactions that the carrier protein undergoes is seen. This starts with the activation of the enzymatic system by chelation of the cation on the luminal side of the membrane. Two enzymatic reactions occur and the ion is transported to the cytoplasm. The carrier protein continues to be involved in other enzymatic reactions to gain energy and return to its original position in order to carry another ion across the membrane.

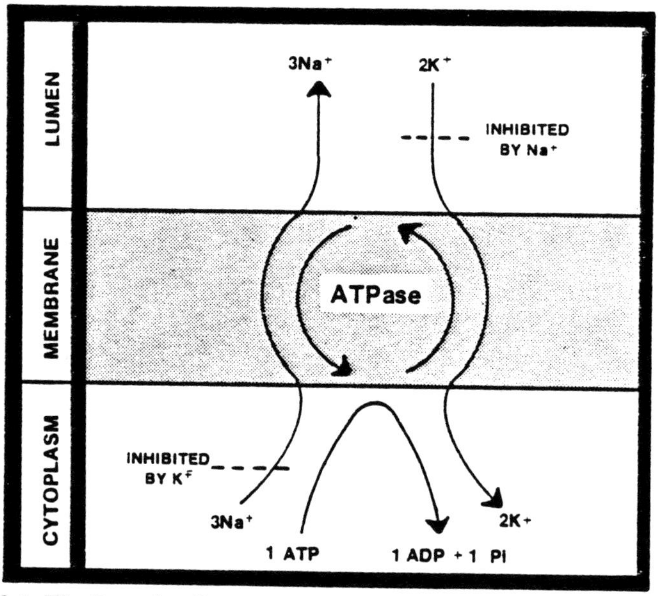

Figure 6-4. The flow of sodium and potassium in the sodium-potassium pump of the membrane as it affects the hydrolysis of ATP.

It has been shown that the cation affinities for the activation of the enzyme pumps on the membrane surfaces are independent of the nature and concentration of the alkali metal ions at the opposite surface.[22,23,24] Thus, Glynn and Karlish[20] concluded that the relation between the movement across the membrane and the cation concentration can be described by the product of a constant and two factors that express the saturation of the cation activation sites on the exterior and the interior membrane surfaces. The affinities for the activating ions at the two surfaces are related to their true-binding affinities and equal to them if there are no competing ions.

The dissociation constant for each metal bound to the carrier protein is different. For example, the dissociation constant for the protein-calcium chelate is 2.6×10^{-3}M.[25]

Henderson[26] has reported that in the enzymatic steps required to traverse the membrane, the distribution of the substrates and

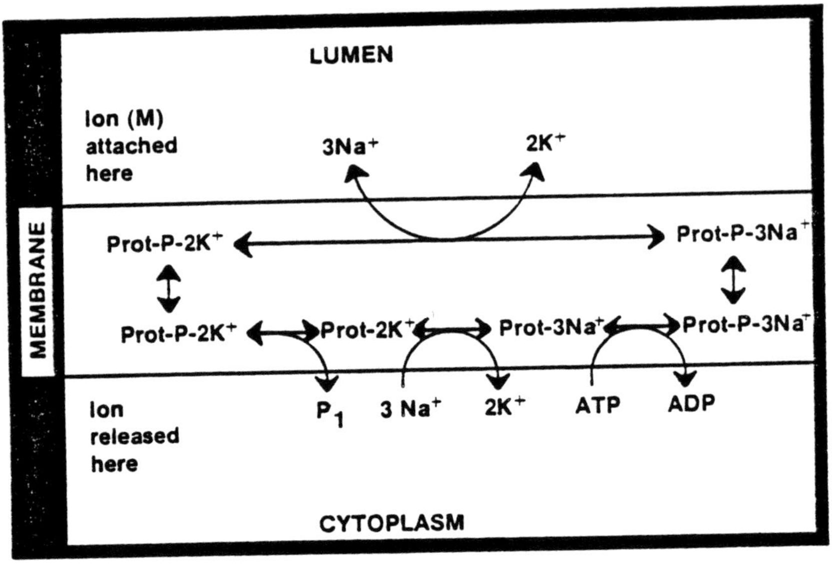

Figure 6-5. The enzymatic reactions involving the carrier protein in the cell membrane.

products of oxidative phosphorylation can result in a pH gradient in the membrane. In other words, as the enzymatic reaction occurs, the pH in the cellular membrane changes, thus favoring the release of the cation at the inner side due to changes in the dissociation constant.

In summary, it appears that the cation is chelated by the integral protein which is part of an enzyme system located on the brush border of the mucosal cell. This binding commences the activation of an enzymatic system called a "pump." Several enzyme reactions, in which the cation is moved from enzyme to enzyme within the system ATP are catalyzed. These hydrolyses cause change in the pH within the membrane which affects the dissociation constant and ultimately favors the release of the cation by the carrier protein to the cytoplasm and its immediate uptake by another protein molecule within the cell. Since there can be up to 5,000 enzyme reactions per minute within a single cell,[27] the movement of the cation from enzyme to enzyme within the single transport system (Figure 6-5) is very rapid. The total reaction stops when the cation is delivered to the interior side of the membrane where the metal is released and rechelated by cytoplasmic proteins, such as apoferritin, in the case of iron.

REFERENCES

1. Kramer, P., *Plant and Soil Water Relationships: A Modern Synthesis* (New York: McGraw Hill Book Co.), 220, 1969.
2. Hobbs, A. and Albers, W., "The Structure of Proteins Involved in Active Membrane Transport," Ann. Rev. Biophys. Bioeng., 9:259, 1980.
3. Singer, S. and Nicolson, G., "The Fluid Mosaic Model of the Structure of Cell Membrane," Science, 175:720, 1972.
4. Pollock, E., "Fine Structural Analysis of Animal Cell Surfaces: Membrane and Cell Surface Topography," Am. Zool., 18:25, 1978.
5. Trier, J. and Madara, J., "Functional Morphology of the Mucosa of the Small Intestine," in Johnson, L., ed., *Physiology of the Gastrointestinal Tract* (New York: Raven Press), V2, 931, 1981.
6. Kowarski, S. and Schachter, D., "Vitamin D-Dependent Particulate, Calcium-Binding Activity and Intestinal Calcium Transport," Am. J. Physiol., 229:1198, 1975.
7. Miller, A., *et al.*, "Isolation of a Vitamin D-Dependent, Calcium-Binding Protein from Brush Borders of Rat Duodenal Mucosa," FEBS Lett., 103:319, 1979.
8. White,, A., *et al., Principles of Biochemistry* (New York: McGraw Hill Book Co.), 296, 1973.
9. Giese, A., *Cell Physiology* (Philadelphia: Saunders College), 396-8, 1973.
10. Robertis, E., *et al., Cell Biology* (Philadelphia: W. B. Saunders Co.), 474, 1975.
11. Giese, op. cit., 407
12. Ibid., 401.
13. White, op. cit., 298.
14. Hobbs, A. and Albers, R., "The Structure of Proteins Involved in Active Membrane Transport," Ann. Rev. Biophys. Bioeng., 9:259, 1980.
15. Knowles, A. and Racker, E., "Properties of a Reconstituted Calcium Pump," J. Biol. Chem., 250:3538, 1975.
16. Ostwald, T. and MacLennan, D., "Isolation of a High Affinity Calcium-Binding Protein from Sarcoplasmic Reticulum," J. Biol. Chem., 249:974, 1974.
17. Miyamoto, H. and Kasai, M., "Asymmetric Distribution of Calcium-Binding Sites of Sarcoplasmic Reticulum," J. Biochem., 85:765, 1979.
18. Martonosi, A., "Calcium Pumps," Fed. Proc., 39:2401, 1980.
19. Boyer, P. and Masahiro, A., "O-Probes of Phosphoenzyme Formation and Cooperativity with Sarcoplasmic Reticulum ATPase," Fed. Proc., 39:2411, 1980.
20. Glynn, I. and Karlish, S., "The Sodium Pump," Ann. Rev. Physiol., 37:1122, 1975.
21. Robertis, op. cit., 478.
22. Hoffman, P. and Tosteson, D., "Active Sodium and Potassium Transport in High Potassium and Low Potassium Sheep Red Cells," J. Gen. Physiol., 58:438, 1971.

23. Garay, R. and Garrahan, P., "The Interaction of Sodium and Potassium with the Sodium Pump in Red Cells," J. Physiol., 321:297, 1973.
24. Garrahan, P. and Garay, R., "A Kinetic Study of the Na Pump in Red Cells. Its Relevance to the Mechanism of Active Transport," Ann. N.Y. Acad. Sci., 1974.
25. Oxender, D., "Membrane Transport," Ann. Rev. Biochem., 41:777, 1972.
26. Henderson, P., "Ion Transport by Energy-Conserving Biological Membranes," Ann. Rev. Micbiol., 25:393, 1971.
27. Ashmead, H., "Metabolic Systems and Their Mineral Activators," paper given at Miller Physician Seminar, Denver, Colorado, December, 1968.
28. Larrson, l. and Rhfeld, J., "A Peptide Resembling COOH-Terminal Tetrapeptide Amide of Gastrin From a New Gastrointestinal Cell Type," Nature, 277:575, 1979.

CHAPTER 7

A SECONDARY INTESTINAL PATHWAY FOR ABSORPTION OF DIPEPTIDE-LIKE AMINO ACID CHELATES

IN the preceding chapter a brief model was developed to describe the absorption of metal ions from the lumen into the mucosal cells of the small intestine. Many investigators have used this or adaptations of this type of active transport model to describe the intestinal uptake of all mineral forms. Unfortunately, this model is descriptive of only those metals that are in an ionic state when presented to the brush border of the mucosal cell. It does not take into account the fact that frequently minerals which are bound to dietary protein fractions are also absorbed without even being ionized in the gut.

Between 1954 and 1956, Mills[1,2,3,4] determined that the greater part of the copper in fresh herbage was not ionizable but instead was in the complexed state. When he fed copper-deficient rats these copper chelates and compared liver storage of copper in copper-deficient rats to liver storage of copper in rats fed copper sulfate, he found both a more rapid response and a greater copper storage in the rats receiving the copper complex.

Presumably the copper from the copper sulfate was ionized in the gut and then principally absorbed by active transport as described in the previous chapter. On the other hand, if the copper complex also required dissociation before mucosal uptake, then its rate of transport to the liver for storage should have been similar to that of the copper from the sulfate. This presumption is based on the fact that both forms of copper ultimately would have been in the same ionic state and therefore dependent on the same carrier mechanism.

But as Mills observed, there was a dissimilarity in the rate of uptake into the liver between the two sources of copper. This suggested two different mechanisms. Mills concluded that copper may be transported through the intestinal mucosa both as ionic copper (from the copper sulfate) and in the form of intact complexes.

A similar observation has been made in the intestinal absorption of iron. As was previously shown, duodenal absorption occurs when the iron is mainly in the ferrous (+2) form,[5] although this is not absolute.[6] Thus, generally speaking, inorganic forms of iron need to be released from conjugation by the gastric juice and other digestive secretions and then reduced to the ferrous state for effective absorption.

The absorption of this inorganic iron is highly sensitive to the environment within the intestine. For example, the inclusion of ascorbic acid with the inorganic iron reduces the iron to the ferrous state and enhances its intestinal absorption through maintaining a lower pH for a longer period of time.[7,8,9,10]

Not only is ionic iron taken up by mucosal cells, but hemoglobin iron is also well absorbed.[11,12,13,14,15] When Hallberg and Solvell included ascorbic acid with the hemoglobin iron, unlike its effect on ionic iron,[7,8,9,10] it had no effect on the absorption of hemoglobin iron. They reported that, "ascorbic acid increases the absorption of both ferrous and ferric iron, but no absorption-promoting effect has been found on the absorption of hemoglobin iron."[16] That suggests a different uptake mechanism for complexed iron which is independent of the ionic iron pathway.

Furthermore, these same investigators found that presence of phytates reduced the intestinal uptake of iron from ferrous sulfate due to the formation of an insoluble iron complex.[17,18] This same phytate had no effect on the uptake of hemoglobin iron. Thus, it appeared that the hemoglobin did not break down and release the iron from the chelate to join with the phytate. If it did, the phytate would have complexed the ionic iron and rendered it insoluble and unavailable too. As shown in the previous chapter, in order for the metal to be absorbed via the ionic pathway, there must be solubility.

In similar experiments by Turnball, *et al.*[19] they concluded, "the most likely explanation of the apparent immunity of hemoglobin iron to the blocking effect of food and of phytate, and to the facilitating effect of ascorbic acid, is that the hemoglobin molecule or an iron-containing portion such as hemin is absorbed intact." Callender, *et al.* have added, "the absorption of iron from hemoglobin clearly involves a more complicated process than the absorption of inorganic iron salts. . . ."[20]

In a series of studies reported by Kuhn, *et al.*,[9] ^{59}Fe absorption was determined in 58 human volunteers ranging between the ages of 11 and 70 years. The iron supplied was that which occurred naturally in wheat, corn, and soybean plus ferritin isolated from rabbit livers. They concluded that in order for the iron from the foods to be absorbed, it had to be reduced and soluble. Their research led them to conclude that all of the iron from the foods tested went through a common absorption pathway in the mucosal cells except ferritin. They found that when iron is chelated to amino acids in the ferritin form, it took a different intestinal absorption pathway. Perhaps the reason that complexed minerals in foods underwent ionic absorption was because in the complexed state their molecular weights were too high. When the molecular weight of a naturally occurring chelate is reduced, it is absorbed via a different pathway than ionic iron.[21,13]

"Simultaneous administration of hemoglobin and an iron salt, labeled with different radio-iron isotopes, showed that there was no isotopic exchange of iron between the two compounds in the gastrointestinal lumen. All of these studies indicate strongly that hemoglobin is absorbed into the mucosal cells as a very stable iron complex, probably an iron-porphyrin complex.

"A number of independent observations indicated that the iron complex was split within the mucosal cells and that hemoglobin iron was absorbed from the mucosal cells into the plasma in the same chemical form as iron administered as a ferrous salt."[16]

Later research by Wheby and Spyker tends to disagree with the concept of total degradation of the iron complex in the mucosal cell.[21] The alternatives are shown in Figure 7-1.

According to their studies, it is possible for the hemoglobin to be digested into heme and globin in the intestinal lumen. The heme may be further degraded releasing the ferrous iron in the lumen. Thus, three forms of iron may be available for mucosal uptake: (1) hemoglobin, (2) heme from digested hemoglobin, and (3) ferrous

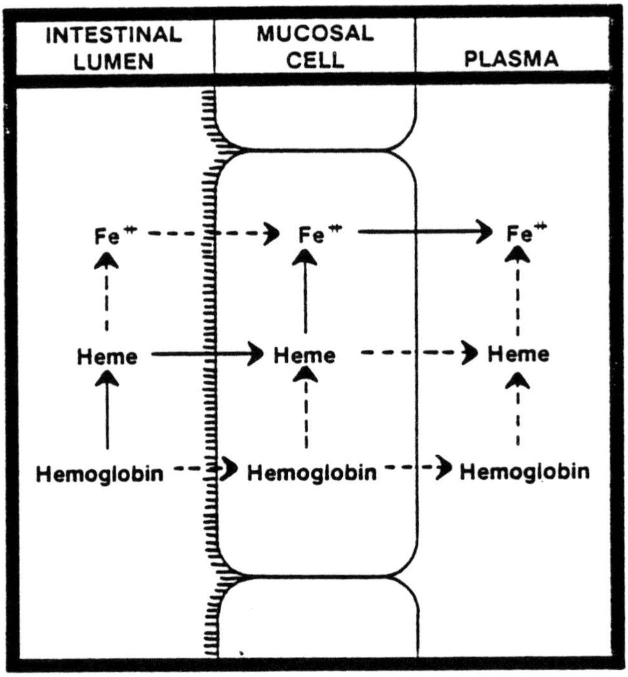

Figure 7-1. Primary and possible alternative pathways for hemoglobin iron absorption. The primary pathway is indicated by solid arrows and possible alternate pathways by broken arrows.

iron from digested heme. This suggests three possible pathways, although our own work cited later rather suggests that the heme uptake is probably greater than that of hemoglobin. Most of the hemoglobin is probably digested into smaller molecules by brush border enzymes prior to absorption.[39,40] As Figure 7-1 shows by the solid line, Wheby and Spyker believe that even though other possibilities exist, heme is the primary source of complexed iron absorbed.

Once the heme has been taken up by the mucosal cell, and as alluded to by Hallberg and Solvell,[16] an enzymatic process within the cell then splits the iron from the porphyrin ring.[6] This iron-heme splitting enzyme appears to be microsomal heme oxygenase.[22] After its release from the heme molecule, the ionic iron is presumably complexed with apoferritin and transported to the plasma as the solid arrow in Figure 7-1 shows.

Nevertheless, Wheby and Spyker are quick to point out that there is some evidence to show that alternative pathways also exist. For example, it was previously shown that a heme complex may be absorbed directly into portal blood.[13] Furthermore, the rapid increase of activity in the carcasses of laboratory animals exposed to ^{59}Fe are suggestive of an alternative path for hemoglobin absorption.

From the brief study of the physiology of the small intestine and particularly the mucosal cell membrane in Chapter 3, it would seem logical that intact hemoglobin would have difficulty in penetrating the cell membrane due to its molecular size. Its digestion in the gut as proposed by Wheby and Spyker would seem reasonable. The ultimate digestion of the iron complex would, of course, be to form a low molecular weight heme molecule having a molecular weight of about 664[38]. The hemoglobin molecule has a molecular weight of about 64,500, while ferritin has a molecular weight of about 500,000.

In a study reported by Layrisse, *et al.*,[23] 2 mg. of ferritin iron were administered alone to twenty human volunteers. Fifteen days later the same people were fed 1 mg. of ferritin irön mixed with raw ground veal liver. The mean absorption from the ferritin iron in the first trial was 1.7%. When it was included with the veal liver, absorption of the iron ferritin increased to 6.6% suggesting digestion of the ferritin and the veal liver and some rechelation of the iron with amino acids from the digested veal liver. Nevertheless, even that was significantly below the 13.2% absorption of the heme iron from the veal liver which had a lower molecular weight than the ferritin.

These investigators pointed out that iron absorption comes from two basic pools. The first is a non-heme pool composed of ferric and ferrous salts, and vegetable iron. There are many characteristics of the iron ferritin that tend to place it in this pool too. The second pool is a heme iron pool formed by hemoglobin and myoglobin.

A similar observation has been made with zinc. Underwood[27] has reported that different protein sources influence the absorption of zinc bound to proteins in rats, pigs, chicks, and poults. Vegetable sources of zinc are not nearly as well absorbed as those from animal sources of zinc. Part of the reduced absorption of zinc from vegetables may be due to the phytates,[28] but the very fact that the vegetable zinc is affected by the phytates is suggestive of total digestion of the vegetable protein-bound zinc and the consequential uptake as the ionic forms of zinc. This would place vegetable zinc in one absorption pool similar to the one outlined by Layrisse, *et al.*[23]

On the other hand, zinc absorption and utilization is enhanced when ingested with casein or liver extracts—both animal proteins.[29] This is suggestive of a second pool for hemoglobin iron.[23]

The very fact that iron or zinc is chelated at a lower stability constant in vegetables and more strongly in the animal proteins, or heme molecules, suggests that it is not the act of chelation that separates the iron or zinc into two absorption pools but rather the chelating ligands and their resulting stability constants.

Conrad, *et al.*[12] have shown that when the heme is purified, the absorption of the iron was less (3%) when compared to iron absorption from heme plus degraded globin (23%). Martinez-Torres and Layrisse have concluded that products of protein degradation appear to be necessary for the absorption of heme iron either from hemoglobin or myoglobin. Furthermore, the protein degradation during the digestion of meat seems to facilitate the absorption of hemoglobin iron even more.[24] Those substances are amino acids and peptides, ligands which carry the iron into the mucosal cells by a different pathway (23%) when compared to ionic iron's absorption (3%).

In the studies[21] of Wheby and Spyker, previously alluded to, problems with the luminal and mucosal degradation of the hemoglobin, iron decreased its absorption rate as a function of time due to decreased polymerization of the hemoglobin iron in the lumen when they added the amino acid histidine. Histidine chelated with the iron increased the absorption.[25]

In work reported by Martinez-Torres and Layrisse,[26] it was shown that several amino acids influenced the uptake of iron from foods. They concluded, however, that cysteine was the only, or principal, end product from the digestion of animal proteins responsible for the increased absorption of iron from foods.

From these reports it appears that there are two pathways for the intestinal absorption of copper, zinc, and iron and probably for other metals. This pathway appears to be dependent upon the chelation of the ion with amino acids from digested protein.

Based on the observation of a two-pathway absorption mechanism, one of which at first appeared to be dependent upon the presence of digested animal proteins, we questioned the requirement of an animal protein. Both the investigations of Wheby and Spyker[21] and Martinez-Torres and Layrisse[26] indicated that it was not so much the animal protein as it was the amino acids derived from the protein that influenced the uptake of the iron. Consequently, we devised a series of *in vitro* experiments designed to test the intestinal uptake of intact amino acid chelates of various metals compared to their inorganic counterparts as metal salts. The chelates were produced from both vegetable and animal sources of amino acids. The experiments were designed to allow both pathways to operate independently, if they both existed in the intestine.

Adult male Sprague-Dawley rats were selected for these experiments. Their mean body weights were all within 15 grams of each other. All were maintained in wire screen bottom cages and fed *ad libitum*, a commercially prepared food (Purina Laboratory Rat Chow). This diet was considered adequately fortified with minerals. It was important that their nutrition be adequate in order to prevent an unusual mineral uptake from the intestines at the time of the experiments due to a deficiency condition.

The techniques discussed by Agar, *et al.*[30] were generally followed since this procedure is acceptable not only for tissue accumulation of amino acids, but minerals as well.

On the day of experimentation, each rat was anesthetized and then sacrificed by cervical dislocation. The intestines were quickly removed and washed three times in a cold Krebs-Ringer's Bicarbonate Buffer (pH 7.4) Solution (called KRB hereafter).

A jejunal segment of the small intestine beginning ten centimeters below the pylorus and continuing for another twenty centimeters was delineated and removed. The remainder of the intestine was discarded.

The twenty centimeter segments were quickly introduced into large petri dishes of buffer solutions maintained at 5°C by placing them on melting ice. While in these petri dishes, the samples were cut into two centimeter segments (ten per sample). Each two centimeter segment was severed longitudinally along the mesenteric line with needle-point scissors and then randomized between animals.

The randomized samples were pre-incubated for one minute in KRB at a temperature of 37°C. At the same time, a mixture of 95.5% oxygen and 4.5% carbon dioxide was bubbled through the solution using fritted glass tubes.

Following pre-incubation, three segment groups were randomly selected and each group placed in a separate incubation beaker into which the above gas mixture was bubbled. Each beaker contained either a copper, iron, magnesium or zinc salt or that same metal as an amino acid chelate in solution.

The chelates were made using ligands derived from whole fish meal (an animal protein), isolated soy protein (a plant protein), or casein from milk (an animal protein). Metal contents varied in the chelates from 10.6% to 11.2% as determined by atomic absorption spectrophotometry. To insure solubility of the chelates prior to the experiment, each chelate was introduced into the presence of the hydrolytic enzyme, thermolysin, (five milligrams per 50 milligrams of chelate) in a metabolic incubator. The mixture was maintained at 37°C and shaken for two hours.

The metal salts as the carbonate (CO_3), the oxide (O_2), and the sulfate (SO_4) were dissolved in thermolysin solutions similar to those that had been used to dissolve the chelates. Each solution had a metal concentration that was equivalent to the chelates as determined by atomic absorption spectrophotometry. All inorganic salts were reagent grade.

Each group of three segments was exposed to 50 micrograms of metal in solution from either one of the salts or chelates for two minutes.

At the conclusion of the two-minute incubation period, the samples were removed from the solutions, washed six times in cold buffered water to remove any external chelate or metal, blotted on filter paper, and placed in numbered pre-weighed foil tares. The samples were dried to a constant weight at 105°C. When dry, they were reweighed prior to analysis. Analysis was done on a Perkin-Elmer 290 Atomic Absorption Spectrophotometer.

The results of the atomic absorption spectrophotometric analysis are shown in Table 7-1. All data are shown as the mean of the three samples in each group.

The values presented in Table 7-1 are the parts per million (ppm) of metal in the dry tissue. Mucosal uptake from the chelated sources exceeded the absorption from equivalent non-chelated inorganic ions with one exception: The carbonate form of iron and magnesium resulted in greater absorption than the chelates from soy protein and whey protein. The animal protein source from fish was

Table 7-1.

INTESTINAL UPTAKE OF DIFFERENT MINERAL FORMS

	CONTROL	ANIMAL PROTEIN CHELATE	PLANT PROTEIN CHELATE	WHEY CHELATE	CARBONATE	SULFATE	OXIDE
COPPER	trace	33	35	17	12	8	11
MAGNESIUM	7	94	57	52	77	36	23
IRON	23	298	130	80	82	78	61
ZINC	14	191	191	126	87	84	66

Data are expressed as ppm of metal in dry mucosal tissue.

absorbed better than the carbonate form.

The differences in uptake between copper chelated with plant protein and copper chelated with the animal protein were deemed insignificant (35:33 ppm).

The control values for all of the minerals were well below the values obtained from those tissue samples incubated in either the chelated or inorganic mineral solution. Mineral uptake by the mucosal cells definitely occurred in the samples exposed to the various mineral forms.

Additional studies using similar as well as different concentrations of metal in the incubation solution were essentially duplicable.[31] From these studies it was obvious that when the metal was chelated with amino acids, its absorption was related to the chelation potential of the ligands from the different protein sources rather than the source of protein, particularly since the quantity of protein available for chelation was held constant.

Differences in the procedures used to form the chelate which were not considered in the other experiments previously discussed, may affect absorption too. In plants, for example, the molecular weight of the dipeptide-like amino acid chelate (ratio of ligand-to-metal of 2:1) should not exceed 1500 if membrane transport of an intact chelate is to take place.[32]

Finally, the amino acid mix within a given ligand may also account for differences in absorption. As indicated above, certain amino acids as well as certain peptide combinations are absorbed more rapidly and at more cellular loci than others.[21,15,33,34] Consequently, chelates made from amino acids derived from proteins deficient in the amino acids that are more readily taken up such as methionine, cysteine, and histidine[33] could result in lower absorption of the metal during a given period, provided of course the chelate is absorbed at the same locus as the peptide/amino acid; the latter provision involves a theory that remains to be proven.

The major point to remember from these experiments is that there was apparent absorption of the chelate into the mucosal cells. Digestion of the dipeptide-like chelates into amino acids and ions prior to their uptake seems very unlikely for several reasons. The first is due to the experimental design. The intestines were washed three times in Krebs-Ringer's Bicarbonate Buffer (pH 7.4) Solution and then pre-incubated for one minute in a similar solution into which oxygen and carbon dioxide were bubbled for equilibration purposes. Presumably any unintegrated external enzymes, which Ugolev reports are the only enzymes in the brush border that hydrolyze peptides,[35] would have been removed by these four washings. Thus, with no external means to break the peptide chelate into its individual components, it had to be absorbed intact into the mucosal cells.

Deductively assuming that the metal was released from its ligands prior to membrane transport, its movement across the mucosal cell membrane would be carrier dependent[36] just as it would be for the ions from the carbonate, the sulfate, and the oxide. If that were the case, then ionic absorption from the salts should have been greater than that from the chelate as a function of time because membrane transport of the metal would have required a second step, the total digestion of the chelate, before ionic absorption could have taken place. This would have delayed the mucosal uptake of the chelate. However, the absorption of the metals from the chelates was increased compared with that from the salts when both were exposed to the intestinal segments for equal periods of time.

Thus, the dipeptide-like amino acid chelated minerals appear to have been absorbed by a different pathway into the mucosa possibly as intact molecules. The existence of two absorption pathways for minerals in the mucosal cells is definitely evident. One allows the chelate to possibly remain intact during absorption. The other requires metallic ionization and chelation on the brush border of the mucosal cell by integral proteins prior to absorption. Later isotope research,[37] which will be discussed in greater detail in a subsequent chapter, confirmed that there was no competition by the metal ion between the two pathways. If chelation precedes mucosal presentation, there is no competition for the integral protein binding sites on the membrane used for ion uptake. This also supports the concept of a second pathway.

These data suggest that the second pathway for mineral absorption in the intestinal mucosa is not as dependent upon a particular protein for intestinal absorption as is on the chelation of the ion with two amino acids to form a small stable complex with a molecular weight similar to that of heme (664).[38] The second pathway becomes operational when such complexes are formed.

REFERENCES

1. Mills, C. F., "Copper Complexes in Grassland Herbage," Biochem, J., 57:603, 1954.
2. Mills, C. F., "Availability of Copper in Freeze-Dried Herbage and Herbage Extracts to Copper-Deficient Rats," Br. J. Nutr., 9:398, 1955.
3. Mills, C. F., "Studies of the Copper Compounds in Aqueous Extracts of Herbage," Biochem, J., 63:187, 1956.
4. Mills, C. F., "The Dietary Availability of Copper in the Form of Naturally Occurring Organic Complexes," Biochem, J., 63:190, 1956.
5. Niccum, W. L., *et al.*, "Use of Ferric and Ferrous Iron in the Prevention of Hypochromic Anemia in Infants," Am. J. Dis. Child., 86:553, 1953.
6. Fritz, J. C., *et al.*, "Biological Availability in Animals of Iron From Common Dietary Sources," J. Agric. Food Chem., 18:647, 1970.
7. Van Campen, D., "Effect of Histidine and Ascorbic Acid on the Absorption and Retention of ^{59}Fe by Iron-Depleted Rats," J. Nutr., 102:165, 1972.
8. Brown, E. and Rother, M., "Studies of the Mechanism of Iron Absorption. II. Influence of Iron Deficiency and Other Conditions on Iron Uptake by Rats," J. Lab. Clin. Med., 62:804, 1963.
9. Kuhn, I., *et al.*, "Observations on the Mechanism of Iron Absorption," Am. J. Clin. Nutr., 21:1184, 1968.
10. Disler, P., *et al.*, "Studies on the Fortification of Cane Sugar with Iron and Ascorbic Acid," Br. J. Nutr., 34:141, 1975.
11. Wheby, N., *et al.*, "Intestinal Absorption of Hemoglobin Iron," Gastroenterology, 58:647, 1970.

12. Conrad, M., *et al.*, "Human Absorption of Hemoglobin Iron," Gastroenterology, 53:5, 1967.
13. Brown, E., *et al.*, "Absorption of Radioiron-Labeled Hemoglobin by Dogs," J. Lab. Clin. Med., 72:58, 1968.
14. Bannerman, R., "Quantitative Aspects of Hemoglobin-Iron Absorption," J. Clin. Med., 65:944, 1965.
15. Weintraub, L., *et al.*, "Absorption of Hemoglobin Iron: The Role of a Heme-Splitting Substance in the Intestinal Mucosa," J. Clin. Invest., 47:531, 1968.
16. Hallberg, L. and Solvell, L., "Absorption of Hemoglobin Iron in Man," Acta. Med. Scand., 181:335, 1967.
17. Hallberg, L. and Solvell, L., "Absorption of a Single Dose of Iron in Man," Acta. Med. Scand., 19:358, 1960.
18. Manis, J. and Schachter, D., "Active Transport of Iron by Intestine: Features of the Two-Step Mechanism," A. J. Physiol., 203:73, 1962.
19. Turnball, A., *et al.*, "Iron Absorption. IV. The Absorption of Hemoglobin Iron," J. Clin. Nutr., 41:1897, 1962.
20. Callender, S., *et al.*, "Absorption of Hemoglobin Iron," Br. J. Haemat, 3:186, 1957.
21. Wheby, M. and Spyker, D., "Hemoglobin Iron Absorption Kenetics in the Iron-Deficient Dog," Am. J. Clin. Nutr., 34:1686, 1981.
22. Raffin, S., *et al.*, "Intestinal Absorption of Hemoglobin Iron-Heme Cleavage by Mucosal Heme Oxygenase," J. Clin. Invest., 54:1344, 1974.
23. Layrisse, M., *et al.*, "Ferritin Iron Absorption in Man," Blood, 45:689, 1975.
24. Martinez-Torres, C. and Layrisse, M., "Iron Absorption from Veal Muscle," Am. J. Clin. Nutr., 24:531, 1971.
25. Conrad, M., "Polymerization and Intraluminal Factors in the Absorption of Hemoglobin Iron," J. Lab. Clin. Med., 69:659, 1966.
26. Martinez-Torres, C. and Layrisse, M., "Effect of Amino Acids on Iron Absorption from a Staple Vegetable Food," Blood, 35:669, 1970.
27. Underwood, E., *Trace Elements in Human and Animal Nutrition* (New York: Academic Press), 207, 1977.
28. O'Dell, B. and Savage, J., "Effect of Phytic Acid on Zinc Availability," Proc. Soc. Exp. Biol. Med., 103:304, 1960.
29. Scott, M. and Seigler, T., "Evidence for Natural Chelates Which Aid in the Utilization of Zinc by Chicks," J. Agri. Food Chem., 11:123, 1963.
30. Agar, W., *et al.*, "The Uptake of Amino Acids by the Intestine," Biochem. et Biophys. Acta., 14:80, 1954.
31. Graff, D., *et al.*, "Absorption of Minerals Compared with Chelates Made From Various Protein Sources Into Rat Jejunal Slices *In Vitro*," Address to Utah Academy Arts, Letters, and Sciences, Salt Lake City, April 1970.
32. Tiffin, L., "Translocation of Micronutrients in Plants," in Dinauer, R., ed., *Micronutrients in Agriculture* (Madison: Soil Science Society), 207, 1972.
33. MacInnis, A., *et al.*, "Specificity of Amino Acid Transport in the Tapeworm Hymenolepis Diminuta and its Rat Host," Rice University Studies, 62:183, 1976.
34. Waldron-Edward, D., "Effects of pH and Counter-Ion on Absorption of

Metal Ions," in Skoryna, S. and Waldron-Edward, D., eds., *Intestinal Absorption of Metal Ions, Trace Elements and Radionuclides* (Oxford: Pergamon Press), 381, 1971.
35. Ugolev,, A., *Physiology and Pathology of Membrane Digestion* (New York: Plenum Press), 1968.
36. Williams, S., *Nutrition and Diet Therapy* (St. Louis: C. V. Mosby Co.), 197, 1977.
37. Fang, S., *et al.*, "Bioavailability of Zinc: Effect of Amino Acid Chelation," in Ashmead, D., ed., *Chelation Mineral Nutrition in Plants, Animals and Man* (Springfield: Charles C Thomas), 137, 1982.
38. Windolz, M., *et al.*, ed., *The Merck Index* (Rahway: Merck & Co.), 606, 1976.
39. Hendrix, T., "The Absorptive Function of the Alimentary Canal," in Mountcastle, V., ed., *Medical Physiology* (St. Louis: C. V. Mosby Co.), V2, 1162, 1974.
40. Radhakrishnan, A., "Intestinal Dipeptidase and Dipeptide Transport in the Monkey and in Man," in Elliott, K. and O'Connor, M., eds., *Peptide Transport and Hydrolysis* (Amsterdam: Elsevier), 37, 50, 56, 1977.

CHAPTER 8

PEPTIDE ABSORPTION IN THE SMALL INTESTINE

IT was shown in the preceeding chapters that the mucosal uptake of metal ions can follow at least two transport mechanisms. The first involves ionization of the metal salt in the gut and the resulting chelation by the integral proteins transversing on the mucosal cell membrane wall. Through a process of active transport the cation is moved from the outside of the cell to the cytoplasm.

In spite of this established and generally accepted model of mineral absorption, it does not account for the absorption of these same minerals as small molecular weight chelates. Luminal hydrolysis of the chelate which would render the metal to its ionic state prior to absorption is not a satisfactory explanation because evidence exists showing a more rapid mucosal uptake of the intact chelate. This supports the concept of a secondary pathway for the minerals.[1,2]

A quick review of chelation with amino acids indicates that generally one amino group from the amino acid forms a complex covalent bond with the mineral. The other donor group can come from the carboxyl group or in the case of some amino acids, the sulfurhydral group. A heterocyclic ring compound with at least five bonds results.[3] For maximum absorption as an intact chelate, the metal is bonded by two ligands producing a double heterocyclic ring which, for all intents and purposes, appears to be much like a dipeptide in the gut. This concept is seen in Figure 8-1 and confirmed by X-ray diffraction[3,4] in Figure 8-2.

The bonding of two amino acid ligands to a metal necessarily

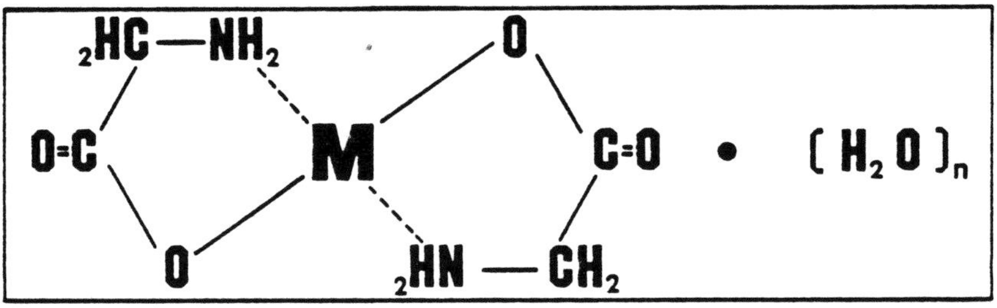

Figure 8-1. Theoretical dipeptide-like amino acid chelate. M represents the mineral.

Figure 8-2. X-ray diffraction showing a dipeptide-like amino acid chelate.

results in a dipeptide-like amino acid chelate. If a true double heterocyclic chelate is formed with the correct coordinate covalent bonding[5] for stability and if those bonds are protected during its production in order to maintain electrically neutral amino acids in the pH of the stomach and small intestines[8] and if the molecular weight of the chelate is under 1500,[6,7] then there would be little opposition to the chelate being absorbed as a dipeptide in the intestine and entering the mucosal cell via the peptide locus.

Payne has written, "When an impermeant substance is chemically linked to a peptide, and the resultant complex can be transported via a peptide permease, the impermeant moiety can thus be 'smuggled' across the cell membrane, and the complex is accordingly called a 'smuggler,'".[9] Fickel and Gilvarg first alluded to this "smuggler" concept in 1973.[10] About the same time Ames, *et al.*, also published their observations concerning the ability of peptides to transport other substances into the cell undetected using the peptide transport mechanism.

Mellander was very specific as to what he felt peptides could "smuggle." In 1947 he began working with casein from human milk.[12] Casein is a hydrolyzed protein of milk being mostly a mixture of peptides. He prepared chelates of calcium with complexed phosphorus. The chelates contained about 10% calcium to ligand and 5% to 6% phosphorus.[13]

In several cases Mellander gave the calcium peptide-like chelates to infants and found very high calcium utilization.[14] Finally, he hypothesized metal chelates of this type can act as physiological carriers for calcium and other metals.

In 1972 Mellander and Folsch described the chemical synthesis of phosphorylated peptides. Generally with this process the resulting peptide bonds close to the serine residues are more enzyme resistant in the gut.[16] This type of phosphorylated peptide, Leu-P-Leu, was tested for absorption in everted hamster jejunum rings *in vitro*. As predicted by Mellander and Folsch, the intestinal hydrolysis of the metal dipeptide complex was considerably slower, and there was evidence of carrier mediated, or active transport, of some of the intact peptide.[17]

In discussing their work with iron incorporated into peptides, Hemmings and Williams complexed labeled ferritin to immunoglobulin.[18] The passage of the intact immunoglobulin across the gut of suckling rats occurs for the first 21 days of life,[19] so this was a period of time in which the dose of labeled ferritin chelated to the immunoglobulin was administered. When blood samples were taken from the offspring, 75% of the ferritin in the serum was no longer precipitable with specific immune serum against ferritin.[18] This demonstrated that 75% of the ferritin complexed to the peptide molecule was absorbed intact with only a 25% hydrolysis. That which was hydrolyzed occurred inside the mucosal cell.

Since chelates with amino acid ligands can act as carriers to transport metals across the cell membrane, and since it has been seen that there is a secondary, non-ionic route for the absorption of minerals when they are chelated to amino acids to form a dipeptide-like chelate, it becomes important to closely examine the absorption of the peptide. Research in the later chapters will be presented as proof that minerals can be absorbed through the peptide absorption sites as part of the dipeptide-like molecule.

The work of MacInnis, *et al.*,[20] shows that cells have specific absorption sites for many amino acids with some overlap in affinity.

The major exception is methionine. It is absorbed at all amino acid absorption loci. Like the amino acids, peptides too have specific absorption sites. Nevertheless, until after the chyme has been mixed with the pancreatic enzymes which are introduced into the lumen following ionic absorption, little intestinal absorption of peptides from the hydrolysis of protein occurs. The digestion of protein is reviewed in Figure 8-3.[21]

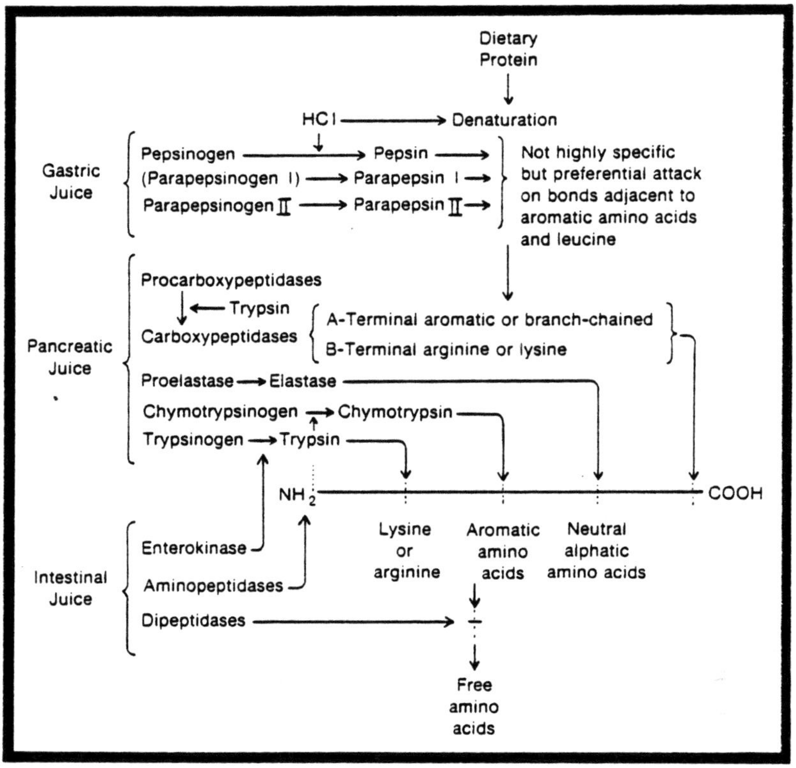

Figure 8-3. The digestion of dietary protein into free amino acids. Digestion starts in the stomach and continues in the intestine.

It is recognized that protein sources not only include food but also intestinal enzymes, mucus, desquamated cells and plasma protein.[22] Nevertheless, for purposes of illustration ingested food will be considered as the only source of protein.

The hydrolysis of this protein commences in the stomach with the action of pepsin. This enzyme attacks the interior peptide bonds of the protein and is most active in the acid pH of the stomach until the pH is raised to about 4. Some amino acids and mostly polypeptides are produced from this digestion.[23]

This digestion is at best incomplete and impaired dramatically as the chyme enters the small intestine with the elevated pH. Once the chyme has passed the duodenum (the predominant site for ion absorption), the proteolytic enzymes of the pancreas are secreted into the intestine where they mix with the chyme. These enzymes (trypsin, chymotrypsin, carboxypeptidase A and B, elastase, ribonuclease, and deoxyribonuclease) break specific peptide bonds. The combined action of these enzymes results in a digest of free amino acids and small polypeptides and dipeptides.[24] At this point, the peptides are in much greater quantities than the amino acids.[25]

The amino acids enter the mucosal cells through several different transport sites that are group specific with some overlap.[20,25] These mechanisms are not available to the peptides. This was first distinguished in 1968 when it was shown that higher plasma concentrations of glycine resulted from the oral administration of glycine peptides compared to equivalent amounts of free glycine.[27] The evidence for distinct and independent carrier mechanisms for small peptides has been summarized by Rubino and Guandalini[28] and is of three main types. First, there is no competition between amino acids and peptides for mucosal uptake, whereas there is competition between peptides. Secondly, in certain cases of congenital defects in amino acid transport systems, these same amino acids are absorbed normally or almost normally when presented to the mucosa as dipeptides. And finally, there is a different type of interaction between the amino acids or the peptides and sodium ions during transport. These findings suggest that peptides are absorbed from the lumen by transport processes that are different from those used by their constituent free amino acids.

The hydrolysis of oligopeptides into amino acids requires the activity of the enzyme group aminopeptidase. "The intestinal epithelium contains high levels of peptidase activity, whereas pancreatic secretion or the intraluminal content of jejunum has trace amounts of this activity. The intestinal epithelium, therefore, appears to be the principal site for the terminal digestion of luminal peptides."[26] Peptides which are hydrolyzed into amino acids by brush border enzymes are then transported into the mucosal cell by the usual amino acid transport mechanism. There may also be a second transport mechanism for these amino acids which has not yet been identified. In that case, the hydrolytic products would not be in equilibrium with the luminal amino acids.[25]

The other possibility for the peptides that are presented to the mucosal cell is that they enter the cell intact. Brush border hydrolysis is still an important principle in this case because molecular size of the peptide will tend to limit its transport across the cell membrane as an intact molecule. In that situation it must be further digested before absorption.

Matthews has indicated that the membrane transport of peptides appears to be dependent upon the molecular size and structural features of the peptide.[29,30] It is possible that very large peptides may utilize the mechanism used to transport whole proteins. On the other hand, lipid-soluble peptides perhaps diffuse through the lipid membrane of the cell. And finally, small peptides which result from protein digestion utilize a carrier-mediated transport system that results in large scale transmembrane transport.

There is good evidence for size discrimination in oligopeptide transport into the cells, and is based in part on the structure of the cell membrane. The peripheral layer in the cell membrane acts as a barrier, or sieve, to the passage of large peptides.[31,32] Although peptides of up to seven amino acid residues have been observed to be absorbed, dipeptides and tripeptides are taken up much more readily, with dipeptides being the most actively transported. For all intents, a tetrapeptide shows no appreciable uptake.[33] Thus from a theoretical point of view, using the amino acid cystine as the example, a pure tripeptide composed of this amino acid would have a molecular weight of 720. 7 (3×240.30) and suggests that this is the maximum molecular size for efficient absorption.[34] Dipeptides are predominantly absorbed intact into mucosal cells according to work done by Silk.[35] For example, it has been shown that glycyl-L-proline is a dipeptide which is not hydrolyzed at the brush border of the mucosal membrane and is transported intact.[28] On the other hand, the work of Andria, *et al.*,[36] indicates that as a general rule most peptides can be hydrolyzed at the brush border. In that case most peptides would have affinity for the amino acid and peptide transport systems. Glycyl-L-phenylalanine is an excellent example. It is hydrolyzed at the brush border but also competes with the above-mentioned glycyl-L-proline at the influx site.

Ugolev, *et al.*,[37] have summarized the evidence for intact peptide absorption. This is seen in Table 8-1.

Table 8-1.

EVIDENCE FOR PEPTIDE TRANSPORT AND INTRACELLULAR HYDROLYSIS OF PEPTIDES[37]

1. It can be demonstrated that there are intact peptides in enterocytes, in the serosal fluid of everted sacs and in the blood.
2. There is active accumulation of poorly hydrolyzed peptides in the enterocyte.
3. There are increased rates of peptide absorption compared to equivalent free amino acids.
4. There is an absence of competition for the transport mechanism between peptides and amino acids and distinct competition between peptides for absorption carriers.
5. There is an absence or reduction of competition for the transport mechanism between amino acid constituents of a peptide.
6. Normal peptide uptake has been observed in the free amino acid transport defects of Hartnup disease and cystinuria.
7. There is an inhibition of active transport and hydrolysis of dipeptides by intact but not homogenized mucosa: (a) in the presence of dinitrophenol; (b) under anoxia.
8. There is an inhibition of free metionine uptake by L-amino acid oxidase (in concentrations of 0.1-0.2 mg/ml), but there is an absence of this effect on the transport of methionine from L-Met-L-Met.
9. Decrease in dipeptide uptake (after acidification of the medium) correlates with reduction in its transport rather than hydrolysis.
10. There is an increase in the rate of dipeptide hydrolysis by mucosal homogenates but not intact mucosa in the presence of Co^{2+}.
11. There are different sites for amino acid and peptide uptake along the small intestine.
12. There is low content of dipeptidase activity in the brush border of enteocytes compared to cystosol.

Peptide absorption is a combination of two processes. First there is brush border hydrolysis of larger peptides into amino acids with uptake of amino acids as described above, and second there is peptide transport into the cell with corresponding intracellular hydrolysis or transport of intact peptides to the blood. It has been estimated that the brush border contains between 5% and 15% of the dipeptidases.[38-41] This suggests that perhaps as much as 85% or more of the dipeptides are transported into the mucosal cells intact.

On the other hand, the work that Peters[40] has reported indicates that about 60% of all of the tripeptides are split on the brush border leaving only 40% to be absorbed into the mucosal cells unhydrolyzed. Almost all of the tetra-, penta- and hexapeptidase activities occur on the brush border. As was noted earlier, most peptides larger than a tripeptide are not absorbed into the mucosal cell.[33] Ugolev, *et al.*, estimate that between 40% and 60% of the peptide bonds are hydrolyzed at the brush border.[37] This is summarized in Table 8-2.[37]

Table 8-2.

EVIDENCE FOR MEMBRANE HYDROLYSIS OF LARGE PEPTIDES AT THE BRUSH BORDER[37]
1. There is increased rates of peptide absorption compared to equivalent free amino acids.
2. There is absence or reduction of competition for the transport mechanism between amino acid constituents of a peptide.
3. There is anoxia-produced inhibition of the active transport of peptides and free amino acids but in the absence of anoxia-produced inhibition of dipeptide hydrolysis by intact and homogenized intestinal mucosa.
4. The presence of peptide hydrolases splitting both dipeptides (up to 20%) and tripeptides (up to 80%) in the isolated brush border fraction has been noted.
6. There is immunological identification of aminopeptidases and disaccharidases on the surface of the brush border vesicles.
7. It is possible to solubilize peptide hydrolysates by papain and Triton X-100 from the surface of the brush border vesicles.
8. There are different sites for amino acid and peptide uptake along the small intestine.
9. There is normal peptide uptake in free amino acid transport defects during Hartnup disease and cystinuria.

Peptidase activity in the villus reflects the existence of two relatively autonomous enzymatic systems (cystol and membrane). A distinct absorption gradient has been established along the length of the intestinal villus. The maximum peptidase activity occurs close to the tip of the villus.[42] Because of this, amino acids are accumulated mainly by the mucosal cells closest to the villus tip,[43] which suggests that as the mucosal cells mature and migrate to the tip of the villus, their brush borders increase their production of peptidase.

"The subcellular distribution of peptidase in the small intestinal mucosa is shown in Table [8-3]. . . . Most of the activity is located in two fractions, the brush border and the cytoplasm. The proportion of enzyme activity in these two fractions differs considerably, depending on the type and chain length of the substrates used. When dipeptidases are used as substrates, less than 10% of the total cellular activity has been found there. When two tetrapeptides were examined, nearly all the activity was in the brush border, with no detectable activity in the cytoplasm."[44]

Cytosal enzymes appear to be different than brush border enzymes.[45] Peptides composed of longer chains than tripeptides were not hydrolyzed by the cystosal peptidase even after prolonged incubation up to ten hours in time. In contrast to the cytosal enzymes, the brush border peptidases were effective in breaking the peptide bonds containing more than three amino acids including a hexapeptide and an actopeptide. Based on the type of peptide hydrolyzed one can conclude that the brush border enzyme is organ-specific while the cytosol is not.[44]

It has been conclusively demonstrated that the peptidase in the brush border protrude from the membrane into the lumen.[46,47,48] Most of the molecule comprising the enzyme, including the hydrolytic site, extends out of the lipid layer. The hydrophobic peptide anchoring the protein of the enzyme in the bilayer of the membrane is only a very small portion of the total enzyme. It is anchored near the amino end of the protein.

Kim[44] has purified two amino-oligopeptidases from the intestinal brush border of the rat. Each had a molecular weight of about 280,000 as determined by gel filtration. They were each composed of two subunits of equal molecular weight. This was determined when the enzymes were electrophoresed under dissociating conditions on SDS-glycoproteins.[44] This is logical since most membrane

Table 8-3.

SUBCELLULAR DISTRIBUTION OF PEPTIDASES IN SMALL INTESTINAL MUCOSA[44]

Fraction	Guinea-pig Gly-Gly-	Leu-Gly-	Gly-Gly-Gly-	Leu-Gly-Gly-	Try-Gly-Gly-	Leu-Ala-	Rat Leu-Leu-Leu-	Ala-Gly-Gly-Gly-
			%				%	
Brush Border and Nuclei	5	6	20	46	63	11	9	89
Mitochondrial Fractions	2	1	7	5	3	3	1	4
Microsomes	1	1	9	5	4	5	2	7
Cytosol	92	91	64	44	30	81	88	0
Recovery (%)	98	90	87	76	87	98	107	97

bound proteins are glycoproteins.[49] Their similar compositions result in common antigenic properties, substrate specificities and kinetic properties.[44]

Kenny[50] indicated that there are five peptidases known to be located in the brush border of the kidney proximal tubules which is very similar to those of the brush border of the intestinal mucosa. All of them are glycoproteins and protrude from the microvillus membrane. "Three of the group, aminopeptidase M, neutral endopeptidase and serine peptidase, are among the major proteins of the membrane, each contributing about 4% of the protein of the microvillus. Since each of the groups possess a different substrate specificity they form a powerful battery of peptide hydrolases capable of degrading oligopeptides to free amino acids and a few dipeptides, of sequence X-Pro,"[51]

Shoaf, *et al.*,[54] have identified seven peptidases in the brush border membrane. At least three of these enzymes are metal dependent for their activation.[50] For example, it has been shown in the hog membrane there are two zinc molecules per aminopeptidase molecule.[52] That applies to both a monopeptidase M and a monopeptidase A. The zinc can be replaced by cobalt. Manganese will activate about 50% of the enzyme while calcium and magnesium have no effect as far as activating the enzyme is concerned.[53] Neutral endopeptidase is also activated by zinc.[50]

In spite of the fact that these enzymes are metal activated, when a mineral is chelated with a small peptide, it inhibits the hydrolytic activity of the brush border enzymes allowing the metal peptide molecule to be absorbed intact. For example, the work of Kim,[52,55] indicates that when zinc is present with the peptide, the brush border enzymes are inhibited from hydrolyzing the peptide. The same is true when 0.5 to 1.0 mM-cobalt, manganese or magnesium is combined with the peptides.[52] The enzymatic activity on the peptides is also blocked by the presence of thiol inhibiting reagents such as Hg^{+2}.[56]

The significance of this observation is that if the metal ion is chelated to two amino acids to form a dipeptide-like structure, and if this dipeptide-like structure functions as a carrier for the metal as indicated earlier,[10-13] then the very presence of the metal in the peptide prevents peptide hydrolysis and release of the cation into the lumen. Instead, the intact peptide "smuggles" the mineral into the mucosal cell through a peptide pathway.

Because the cation is in a chelated state from the time of ingestion, passage through the stomach, and beyond the ionic absorption sites in the intestine, it has not been free to enter into the multitude of chemical reactions so common with cations derived from ingestion of soluble salts. The result is a much higher absorption rate of the mineral that has been chelated to the amino acids.

Mellander has successfully demonstrated this to be a fact when he prepared a 10% calcium-peptide chelate.[13] Such a chelate was found to be enzyme resistant[16] just as was the previously mentioned mineral-peptide chelates. When this calcium chelate was given *in vivo* (to infants), a very high calcium utilization was noted.[14] It was concluded that under these conditions peptides acted as physiological carriers for the calcium as well as other metals.[15]

If the peptide is hydrolyzed, Figure 8-4 summarizes the possibilities for the ultimate digestion of the substrate.[57]

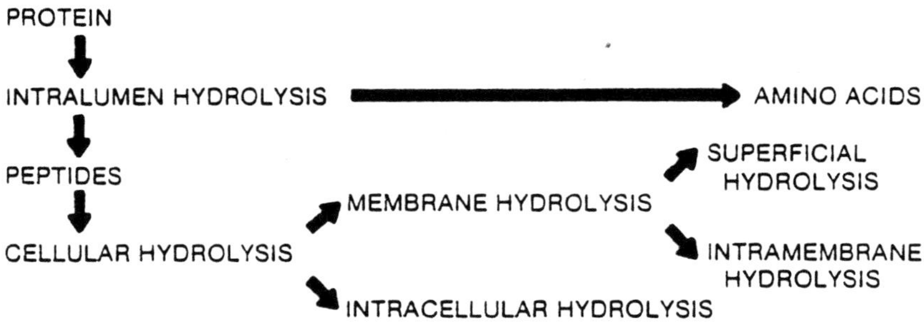

Figure 8-4. Hydrolysis of a peptide.[57]

Intralumen hydrolysis is that digestion of the protein that starts in the stomach and continues in the intestinal lumen. Free amino acids and small peptides are produced. The amino acids and some di- and tripeptides are absorbed. The remaining larger peptides are then subjected to mucosal cell membrane hydrolysis.

Membrane hydrolysis can be further subdivided into two types based on geography. Superficial hydrolysis is the breaking down of the peptides with the corresponding release of amino acids into the luminal solution. These amino acids behave as if they were introduced into the lumen as free amino acids. The intramembrane hydrolysis can be prevented when the peptides are used as ligands for chelating cations, but there is no inhibition on their uptake as intact peptides.

Intracellular hydrolysis is the further digestion of peptides to amino acids after absorption into the mucosal cell. This can occur deep in the plasma membrane, in the cytosol, or some other subcellular organelle. Intracellular hydrolysis will be examined in greater detail below.

It is important to realize that brush border dipeptidase activity is not an essential requirement for uptake of small peptides. Although earlier studies suggested the contrary,[58] it has been concluded that dipeptidase activity does not play any direct role in their transport across the mucosal cell membrane as determined by work done by Radhakrishman.[25]

The transport of free amino acids into the mucosal cells from peptide digestion may involve the sodium-potassium pump energy dependent carriers with some specificity for neutral, basic, and acidic classes of amino acids.[59] The carrier for these amino acids does not differentiate between D and L configurations of the amino acids it accepts.[60]

There are, however, differences in the transport mechanism of free amino acids and peptides. For example, there is an absence of competition between the two for absorption sites.[61] In fact, mucosal cell uptake rates of the peptides tend to be higher than that of the amino acids. "The rapidity with which peptides may be transported, in comparison to the transport rates of the equivalent-free amino acids, is a particularly stiking feature of peptide transport. . . . Peptides may be taken up over 10 times more rapidly than their constituent amino acids in the free form . . . the kinetic characteristics of transport of peptides and transport of their constituent amino acids are different, and it has been suggested that in the intestine, peptide transport is characterized by higher Vmax values than amino acid transport."[62]

Silk has added, "The finding of more rapid absorption of amino acid residues from peptide than from equivalent equimolar amino acid solutions cannot be explained if complete luminal or brush border hydrolysis of peptides precedes uptake of liberated free amino acids. If this were the case, at the best, absorption rates of amino acid residues from the two solutions would be the same. In addition, if complete luminal or surface dipeptide hydrolysis was prerequisite for absorption, the normal differential between absorption of the constituent free amino acids would have been expected. This was clearly not the case. These findings, therefore, suggest that the dipeptides and tripeptides studied are absorbed from the lumen of the normal human small intestine by transport processes that are separate from those used for absorption of their constituent free amino acids".[35]

Small peptides, such as dipeptides and tripeptides, utilize a high capacity carrier-mediated transport mechanism which allows transmembrane transport on a massive scale. This carrier mechanism which is selective for dipeptides and tripeptides is located in the brush border of the intestinal mucosal cell.[63] It was noted above that the transport process for amino acids and peptides is different.[35] So, too, is the transport mechanism. It is distinct and independent of the transport mechanism of amino acids.[28]

This mechanism for the transport of the peptides into the mucosal cell is an active transport process. Ugolev, *et al.*, have noted that it will allow the accumulation into the cell of those peptides that are not hydrolyzed at the brush border against an electrochemical gradient.[37] Whereas the active transport of amino acids into the cell may utilize a sodium-potassium pump,[59] the active transport of peptides does not. During transport, peptides show a different type of interaction with sodium ions compared to amino acids.[64,65] In work done by Sigrist-Nelson it was shown that the rate of uptake of glycyl-L-leucine was not affected any differently by the presence of either sodium or potassium, whereas these same amino acids in the free form were.[66]

The attachment of the amino group on one of the amino acids composing the peptide to the carrier molecule and the peptide's movement across the membrane may require the consecutive action of six enzymes known to be in the intestinal mucosal cells. Meister proposed the following general mechanism for peptide transport[67-70] which is seen in Figure 8-5.

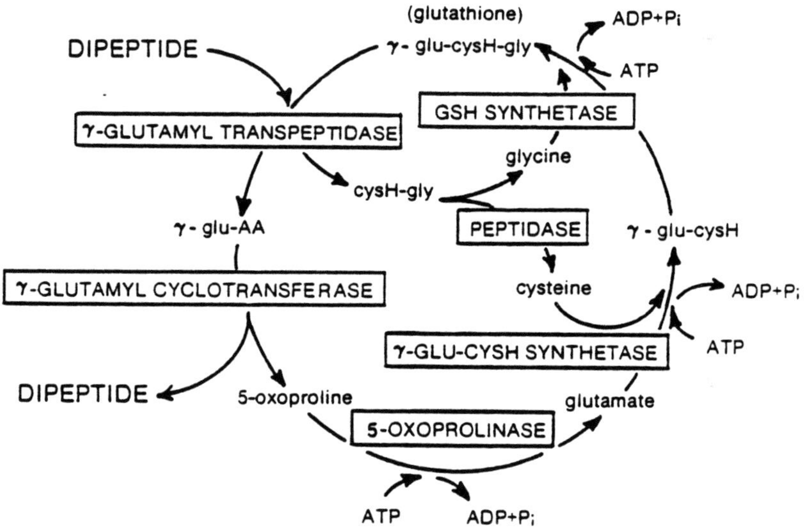

Figure 8-5. The active transport of dipeptides into the mucosal cell.

Glutathione, a tripeptide, is a very abundant intracellular thiol and intracellular γ-glutamyl compound and appears to be the most abundant sulphydryl compound in cells.[73] Since γ-glutamyl transpeptidase is membrane bound and glutathione exists only in the cytoplasm of the mucosal cell, there is as yet an unidentified mechanism that moves the glutathione through the membrane to interact with the transpeptidase.[72] The enzymatic breakdown of the glutathione results in the movement of the peptide from the exterior side of the mucosal cell membrane to the cytoplasm of the cell. At various enzymatic steps the uptake and release of the peptide is accomplished as suggested in Figure 8-6.

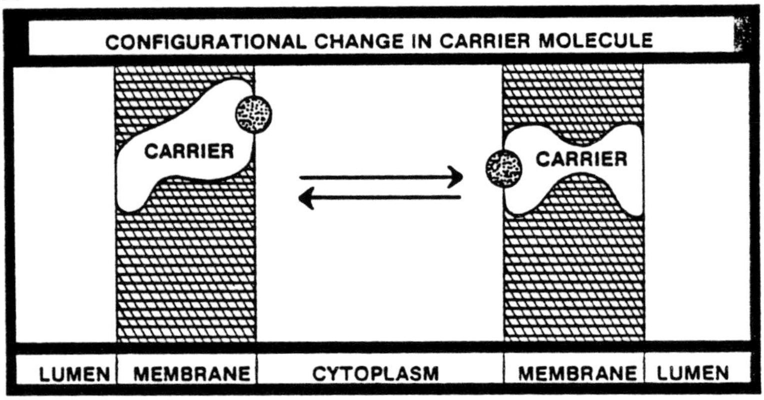

Figure 8-6. Carrier mediated movement of peptides across the cell membrane.

Thus, the "gamma-glutamyl transpeptidase may function in the translocation, and gamma-glutamylcysteine synthetase and glutathione synthetase may catalyze energy requiring recovery steps in transport."[74] As Figure 8-5 shows, there are six step reactions involved. The first reaction is:

$$\text{Glutathione} + \text{peptide} \underset{\text{cylotransferase}}{\overset{\gamma\text{-glutamyl}}{\rightleftharpoons}} \gamma\text{-glutamyl dipeptide} + \text{cysteinyl-glycine}$$

The peptide is attached to the glutathione at the negatively charged glycine moiety, next to the cysteine, their positions in the integral protein that extends into the lumen being suggested in Figure 3-11. The γ-glutamyl-transpeptidase enzyme then catalyzes the transfer of the γ-glutamyl moiety of glutathione to a dipeptide acceptor to form a γ-glutamyl dipeptide compound and cysteinyl-glycine.

The second reaction is:

$$\text{Cysteinylglycine} + H_2O \xrightarrow{\text{Peptidase}} \text{cysteine} + \text{glycine}$$

The cysteinyl-glycine is cleaved by a dipeptidase enzyme. The products of this reaction, cysteine and glycine, are used later to reform glutathione.

Returning to the γ-glutamyl-peptide, it is involved in step three.

γ-glutamyl
L-γ-glutamyl-dipeptide → 5-oxy-L-proline + L-dipeptide.
cyclotransferase

The γ-glutamyl-dipeptide is converted to 5-oxoproline (pyroglutamate or pyrrolidone carboxylate) and the dipeptide by the cleaving action of the enzyme γ-glutamylcyclotransferase. At this point the dipeptide is now inside the cell. It was transported as a γ-glutamyl derivative, after which the γ-glutamyl moiety is removed by cyclization or hydrolysis. It has been moved from the membrane exterior to the cytoplasm with these two enzymatic reactions.

At this point, the active transport of the dipeptide across the membrane has been completed. The remaining reactions are involved in reconstructing glutathione so that the active transport of the next peptide can occur.

The fourth reaction is:

5-oxoprolinase
5-oxo-L-proline + ATP + $2H_2O$ → L-glutamate + ADP + Pi.
Mg^{+2} (Mn^{2+}), $K^{+(NHY)+}$

The 5-oxo-L-proline is not an artifact, but a quantitatively significant metabolite. The reaction is catalyzed by the enzyme 5-oxoprolinase and activated by magnesium, manganese, and potassium ions. It cleaves the internal peptide bond of 5-oxo-L-proline. This is the only known instance in which cleavage of a peptide bond requires energy.

The fifth reaction is:

γ-glutamylcysteine
L-glutamate + L-cysteine + ATP → L-γ-glutamylcysteine + ADP + Pi.
Synthetase

The cysteine was derived from the dipeptidase activity of the cysteinyl-glycine product from reaction one. The glutamate binds noncovalently to γ-glutamyl-cysteine synthetase, attaching to the glutamate and cysteine binding sites of this enzyme.

The glycine, the other product from this dipeptidase activity is involved in the sixth reaction:

glutathione synthetase
L-γ-glutamylcysteine + glycine + ATP → glutathione + ADP + Pi.

Both of these two reactions are ATP-dependent reactions and catalyzed by γ-glutamyl-cysteine synthetase and glutathione synthetase respectively. Both are irreversible reactions. At this point the glutathione has been reformed and is ready to carry another dipeptide across the mucosal cell membrane.[79,67,68,71]

"The dipeptide uptake system includes among its structural requirements a terminal NH_2 group, a terminal COOH group, and a peptide bond. The oligopeptide system is distinguished by its lack of requirement for the terminal COOH group, which gives it the ability to accept peptides of several (up to seven) amino acid residues; probably it could accept larger peptides than this if they were not prevented from reaching it by the limitation in molecular size imposed by the pores in the cell wall surrounding the plasma membrane. Dipeptides can utilize the oligopeptide system but oligopeptides cannot utilize the dipeptide system."[33] While most of the work on bonding has been done in *E. coli* bacteria, it is believed that the same principles probably apply in the mucosal cells, except mucosal cells will generally not absorb tetrapeptides or larger whereas bacteria will.

As stated above, both dipeptide and oligopeptide transport require the presence of a free protonated N-terminal γ-amino group. It must be adjacent to the peptide bond. When the positive charge on the amino group is removed, the peptide is unable to be absorbed. It is not known whether the peptide enters the cell membrane at the C-terminal or the N-terminal. The fact that the N-terminal requirement is absolute implies that this end is oriented towards the membrane.[9]

The requirement of the free C-terminal γ-carboxyl group is absolute for uptake of dipeptides. Removal of this group prevents binding and consequently transport. As stated above, an oligopeptide has no absolute requirement for the COOH group. This allows the oligopeptide to smuggle a substance, such as a metal ion, across the cell membrane by attaching it to the carboxyl group. "When an impermeant substance is chemically linked to a peptide, and the resultant complex can be transported via a peptide permease, the impermeant moiety can thus be 'smuggled' across the cell membrane and the complex is accordingly called a 'smuggler'".[9] There is some evidence that the presence of a D-isomer at the carbon terminus of a tripeptide does not prevent its transport.[9]

It is the requirement of the carboxyl group by the dipeptide transport system and its non-requirement by the oligopeptide transport mechanism that suggest two mucosal carrier systems for peptides.[77] Dipeptides can be taken up by the oligopeptide system, but the converse is not true.[76] The pathway used for peptide absorption can also be used by amino acids.[78] Because of a lack of competition between the amino acids and peptides, in all probability the amino acids utilize the peptide absorption pathway only when the amino acid pathway is saturated or when the peptide uptake mechanism is not occupied with the absorption of peptides.

Figure 8-7 is a summary of the structural requirements for peptide transport. It represents a dipeptide and the N-terminal portion of an oligopeptide.

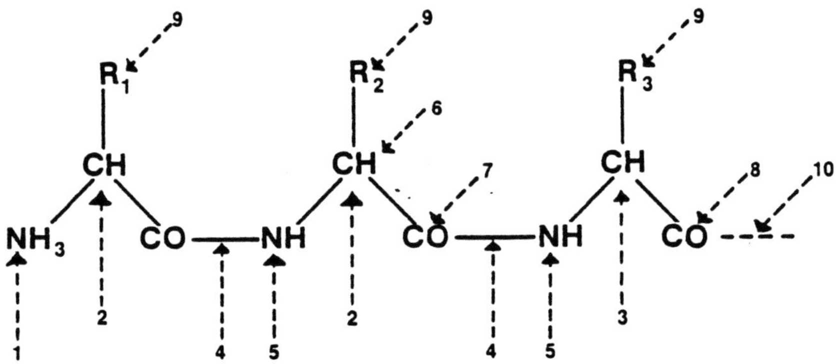

Figure 8-7. The structural requirements for peptide transport.[9]

"Numbers refer to the following features: (1) the γ-amino group should preferably be free but may be monosubstituted providing a positive charge is retained, e.g. monoalkyl; acyl or dralkyl substituents prevent transport; (2) L-isomers specified; (3) in oligopeptides this residue (and perhaps other towards the C-terminus) may be D-isomer; (4) γ linkage specified; (5) nitrogen atom may be methylated; (6) hydrogen not specified; (7) free C-terminal-γ-carboxyl specified for dipeptide transport but not (8) for oligopeptides in which it may be absent or substituted; (9) no absolute specificity for side chains, may be natural residues, synthetic or substituted, but their nature may affect transport affinity and/or rate; possibility of multiple transport systems; (10) large size limits oligopeptide transport."[9]

Peptides are probably absorbed in greater quantities in the ileum than in the jejunum or duodenum of the small intestine.[53] Nevertheless, when villus tip cells and crypt cells of rat jejunal mucosal cells were measured forenzymes of the γ-glutamyl cycle, it was found that the tip cells had much higher γ-glutamyl transpeptidase activities than the crypt cells.[75]

Once the peptide has traversed the mucosal cell membrane, it may undergo intracellular hydrolysis into amino acids. The hydrolysis can take place in the cytosol, in the subcellular organelles, or just below the membrane of the cell in the terminal web. Wacker and Semenza have estimated that approximately 90% of the total aminopeptidase activity of the small intestine is located within the intestinal cell.[79] This is logical since the majority of the peptides that are absorbed are in the dipeptide state. Furthermore, it has been shown that the dipeptidases are responsible for the hydrolysis of most peptides in the cell.[80] There is only a limited number of dipeptidase enzymes in the cytoplasm to hydrolyze a range of over 400 possible dipeptides. Consequently, they have a broad substrate specificity.[25] Table 8-4 lists those enzymes that have been isolated and identified.[80]

Table 8-4.

SUBCELULAR DISTRIBUTION OF ISOLATED AND CHARACTERIZED PEPTIDASES IN THE ENTEROCYTE[80]

Cytosol	Glycylleucine dipeptidase (EC 3.4.13.2)
	Glycylglycine dipeptidase (EC 3.4.13.1)
	Proline dipeptidase (EC 3.4.13.9)
	Aminotripeptidase
Lysosomes	Cathepsins A, B, C and D
Brush Border	Aspartate aminopeptidase (EC 3.4.11.7)
	Glycylprolyl dipeptidylpeptidase
	γ-Glutamyl transpeptidase (EC 2.3.2.2)
	Microsomal aminopeptidase (EC 3.4.11.2)
	Aminotripeptidase

This does not mean that tripeptides cannot be hydrolyzed intracellularly when absorbed intact. While there are over 8,000 possible combinations,[25] Rhodes, *et al.*, showed data that strongly favor intracellular digestion of tripeptides.[53] This is also seen in Table 8-4. Aminotripeptidase is found in the cytosol.

One of the cytosol enzymes has a molecular weight of 107,000 with a single dimeric subunit.[25] Glycylleucine dipeptidase, which is responsible for most of the dipeptidase activity within the mucosal cell, has a molecular weight of 115,000. Proline dipeptidase, a cellular enzyme capable of splitting dipeptides rich in proline or hydroxproline, has a molecular weight of 113,000 with two subunits whose molecular weights are about 53,000.[80] Finally, the aminopeptidase has three subunits. Its total molecular weight is 140,000 to 160,000.[81]

Both glycylleucine dipeptidase and proline dipeptidase belong to the SH enzymes.[82,83] As such they are activated by a specific metal. It is suggested this metal is probably zinc.[80] Each enzyme has two active sites which are not cooperative in their function.[80] There is no indication that the other peptidases found in the cytosol are also activated by a cation. Due to observations made in regards to brush border enzymes and the two dipeptidases mentioned above, it is likely that the remaining cytosol enzymes will be found to be metal dependent, too.

In spite of this cation dependency, the inclusion of certain metals will inactivate their enzymatic action just as these metal ions affect the brush border enzymes. For example, the presence of either zinc[55] or mercury[80] very effectively inhibits the cytoplasmic enzymes. This observation suggests that when a metal chelate is formed with amino acids and takes on the characteristics of a peptide, it not only survives the enzymatic activity of the brush border but also the cytosol enzymes.

That the peptide can avoid hydrolysis and be absorbed into the portal blood as an intact molecule has been proven. "In most cases only free amino acids appear in the portal blood, but peptides may cross the membrane of the intestinal brush border intact; they are hydrolyzed intracellularly, or during transport itself, or, more rarely, not at all. The nature of the peptide, the animal species, and the conditions applied (in particular incubation time, peptide concentration, etc.) may be decisive for the way a given peptide is handled."[79] The fact that dipeptides, in particular, may enter the portal blood is not a reason for alarm. All tissues of the body are equipped to handle dipeptidase activity[84] which suggest that under normal *in vivo* conditions the passage of small intact peptides through the mucosal cell to the blood is a frequent occurrence. And in fact, there is some evidence to suggest that peptides are translocated from one part of the body to another.[84] If this is true, then it is certainly possible that an intact chelate having the same general characteristics of a dipeptide may also be translocated.

The research of Conrad, *et all.*,[85] summarizes the concept of metals chelated with amino acids being absorbed into the mucosal cell through the peptide pathway. Using guinea pigs as the laboratory animal for the experiments, ^{59}iron as hemoglobin was administered. Measurements were made at three hours and again at seven days after the oral dose to determine the percentage of radioiron absorbed by the guinea pigs. In this sense, the iron within the hemoglobin molecule is a component of a large peptide, and even though it is a dietary source of iron, the body reacts to it as it would any peptide.

The study[85] showed that the first reaction was to begin digesting the hemoglobin molecule. Most of the hydrolysis occurred in the duodenum with the introduction of the duodenal enzymes. While hemoglobin has a molecular weight of 67,000 and was too large to be absorbed intact, the iron chelate product of its degradation was within the acceptable range. The resulting metalloporphyrin was a small heme molecule with a molecular weight of 650. This is within the range of a dipeptide, and mucosal absorption is related to molecular size, not configuration of the heme molecule.

The brush border enzyme did not digest the heme further. It was absorbed intact into the mucosal cell. Presumably, the presence of the iron within that heme molecule inhibited the peptidase activity on the cell membrane.

Once absorbed, the heme molecule was transported to the basement membrane of the mucosal cell and absorbed intact into the portal blood. Again, in spite of the very high dipeptidase activity within the cytosol of the mucosal cell, the iron-containing peptide survived further hydrolysis. It is quite possible this was again due to the presence of the iron in the molecule.

Additional work by these same investigators showed that the high level of iron absorption is related to its chelation with the amino acids and being a part of the peptide molecule. Its absorption is much higher than similar quantities of inorganic iron. These data suggest that a secondary pathway is being utilized for the absorption of iron when it is part of a peptide molecule. That pathway is a peptide absorption pathway.

REFERENCES

1. Graff, D., *et al.*, "Absorption of Minerals Compared with Chelates Made from Various Protein Sources *In Vitro*," Address to Utah Academy of Arts, Letters and Sciences, Salt Lake City, April, 1970.
2. Ashmead, H., "Tissue Transport of Trace Minerals," J. of App. Nutri., 22:42, 1970.
3. Mellor, D., "Historical Background and Fundamental Concepts," in Dwyer, F. and Mellor, D., eds., *Chelating Agents and Metal Chelates* (New York: Academic Press), 1, 1964.
4. Dalley, K., and Johnson, B., "X-Ray Defraction of Chelates," Unpublished, 1979.
5. Johnson, B., "Infrared Spectrometric Identification of the Albion Chelate," Unpublished, 1978.
6. Tiffin, L., "Translocation of Micronutrients in Plants," in Dinauer, R., ed., *Micronutrients in Agriculture* (Madison: Soil Science Society), 207, 1972.
7. Johnson, B., "Physical Characteristics of Albion Chelates Versus other Chelates," Unpublished, 1980.
8. White, A., *et al.*, *Principles of Biochemistry* (New York: McGraw Hill), 102, 1973.
9. Payne, J. W., "Transport and Hydrolysis of Peptides by Microorganisms," in Elliott, L. and O'Connor, M., eds., *Peptide Transport and Hydrolysis* (Amsterdam: Elsevier), 307, 310, 320, 329, 1977.
10. Fickel, T. and Gilvarg, C., "Transport of Impermeant Substances in *E. Coli* by Way of Oligopeptide Permease," Nat. New. Biol., 241:162, 1973.
11. Ames, B., *et al.*, "Illicit Transport, the Oligopeptide Permease," Proc. Natl. Acad. Sci., 70:456, 1973.
12. Mellander, O., "On Chemical and Nutritional Differences Between Casein from Human and Cow's Milk," Acta. Soc. Med. Upsal., 52:107, 1947.
13. Mellander, O., "The Nutritional Significance of Some Peptides," Voeding, 16:336, 1955.
14. Mellander, O., "Phosphorylated Peptides Ca Absorption in Infants," Acta. Soc. Med. Upsal., 55:247, 1950.
15. Mellander, O., "Protein Quality," Nutr. Rev., 13:161, 1955.
16. Mellander, O., and Folsch, G., "Enzyme Resistant and Metal Binding of Phosphorylated Peptides," in Bigwood, E., ed., *International Encyclopedia of Food and Nutrition* (Oxford: Pergamon Press), VII; 569, 1972.
17. Mellander, O., "Discussion," in Elliott, K. and O'Connor, M., eds., *Peptide*

Transport and Hydrolysis (Amsterdam: Elsevier), 295, 1977.

18. Hemmings, W. and Williams, E., "Transport of Ferritin Across the Ileum of Suckling Rats," Int. Res. Common. Syst. Med. Sci., 3:215, 1975.
19. Brambell, F., "The Transmission of Passive Immunity for Mother to Young," in *Frontiers of Biology* (Amsterdam: North Holland), Vol. 18, 1970.
20. MacInnis, A. *et al.*, "Specificity of Amino Acid Transport in the Tapeworm Hymenolepis Diminuta and its Rat Host," Rice University Studies, 62:182, 1977.
21. Munro, H., "An Introduction to Biochemical Aspects of Protein Metabolism," in Munro, H. and Allison, B., eds., *Mammalian Protein Metabolism* (New York: Academic Press), VI, 35, 1969.
22. Nasset, K. *et al.*, "The Digestion of Proteins *In Vivo*," J. Nutr., 56:83, 1955.
23. Code, C., ed., *Handbook of Physiology* (Baltimore: Williams & Wilkins Co.), V5, 346, 1968.
24. Hendrix, T., "The Absorptive Function of the Alimentary Canal," in Mountcastle, V., ed., *Medical Physiology* (St. Louis: C. V. Mosby Co.), V2, 1162, 1974.
25. Radhakrishnan, A., "Intestinal Dipeptidase and Dipeptide Transport inthe Monkey and in Man," in Elliott, K. and O'Connor, M., eds., *Peptide Transport and Hydrolysis* (Amsterdam: Elsevier), 37, 50, 56, 1977.
26. Adibi, S. and Kim, Y., "Peptide Absorption and Hydrolysis," in Johnson, L., ed., *Physiology of the Gastrointestinal Tract* (New York: Raven Press), V2, 1078, 1982.
27. Craft, I., *et al.*, "Absorption and Malabsorption of Glycine and Glycine Peptides in Man," Gut., 9:425, 1968.
28. Rubino, A. and Guandalini, S., "Dipeptide Transport in the Intestinal Mucosa of Developing Rabbits," in Elliott, K. and O'Connor, M., eds., *Peptide Transport and Hydrolysis* (Amsterdam: Elsevier), 61, 1977.
29. Matthews, D., "Absorption of Peptides by Mammalian Intestine," in Matthews, D. and Payne, J., eds., *Peptide Transport in Protein Nutrition* (Amsterdam: North Holland), 61-146, 1975.
30. Matthews, D., "Intestinal Absorption of Peptides," Physiol. Rev., 55:537, 1975.
31. Payne, J., "Transport and Hydrolysis of Peptides by Microorganisms," in Elliott, K. and O'Connor, M., eds., *Peptide Transport and Hydrolysis* (Amsterdam: Elsevier), 309, 1977.
32. Matthews, D., and Payne, J., "Peptides in the Nutrition of Microorganisms and Peptides in Relation to Animal Nutrition," in Matthews, D. and Payne, J., eds., *Peptide Transport in Protein Nutrition* (Amsterdam: North Holland), 1-60, 1975.
33. Burston, D., *et al.*, "Mechanisms of Peptide Transport," in Elliott, K. and O'Connor, M., eds., *Peptide Transport and Hydrolysis* (Amsterdam: Elsevier), 79-81, 1977.
34. Windholz, M., ed., *Merch Index* (Rahway: Merck & Co.), 2778, 1976.

35. Silk, D., "Amino Acid and Peptide Absorption in Man," in Elliott, K. and O'Connor, M., eds., *Peptide Transport and Hydrolysis* (Amsterdam: Elsevier), 16, 20, 1977.
36. Andria, D., *et al.*, "Glutamyl-B-Naphthylamide Hydrolase of Rabbit Small Intestine. Localization in the Brush Border and Separation From Other Brush Border Peptidases," Biochem. Biophys. Acta., 419:42, 1976.
37. Ugolev, A., *et al.*, "Membrane and Intracellular Hydrolysis of Peptides: Differentiation, Role and Interrelations With Transport," in Elliott, K. and O'Connor, M., eds., *Peptide Transport and Hydrolysis* (Amsterdam: Elsevier), 221-238, 1977.
38. Rhodes, J., *et al.*, "Studies on the Organization of the Brush Border in Intestinal Epithelial Cells. IV Aminopeptidase Activity in Microvillus Membranes of Hamster Intestinal Brush Borders," Biochem. Biophys. Acta., 135:959, 1967.
39. Peters, T., "Subcellular Fractionation of the Interocyte With Special Reference to Peptide Hydrolysis," in Ciba Found Symp. *Peptide Transport in Bacteria and Mammalian Gut* (Amsterdam: Associated Scientific Publishers), 107-115, 1972
40. Peters, T., "The Subcellular Location of Intestinal Peptide Hydrolases," Matthews, D., "Absorption of Peptides by Mammalian Intestine," in Matthews, D. and Payne, J., eds., *Peptide Transport in Protein Nutrition* (Amsterdam: North Holland), 243, 267, 1975.
41. Fujita, M., *et al.*, "Oligopeptidases of Brush Border Membranes of Rat Small Intestinal Mucosal Cells," J. Physiol. 227:377, 1972.
42. Cornell, J. & Meister, A., "Glutathione and Glutamyl Cycle Enzymes in Crypt and Villus Tip Cells of Rat Jejunal Mucosa," Proc. Natl. Acad. Sci., 73:420, 1976.
43. Kinter, W. and Wilson, T., "Autoradiographic Study of Sugar and Amino Acid Absorption by Everted Sacs of Hamster Intestine," J. Cell. Biol., 25:19, 1965.
44. Kim, Y., "Intestinal Mucosal Hydrolysis of Proteins and Peptides," in Elliott, K. and O'Connor, M., eds., *Peptide Transport and Hydrolysis* (Amsterdam: Elsevier), 154-159, 1977.
45. Kim, Y., *et al.*, "Specifities of Peptide Hydrolases in Brush Border and Cytosol Fractions of Rat Small Intestine," Biochem. Biophys. Acta., 370:283, 1974.
46. Maroux, S. and Loubard, D., "On the Hydrophobic Part of Aminopeptidase and Maltases Which Bind the Enzyme to the Intestinal Brush Border Membrane," Biochem. Biophys. Acta., 419:189, 1975.
47. Louvard, D., *et al.*, "Topological Studies on the Hydrolases Bound to the Intestinal Brush Border Membrane. II. Interactions of Free and Bound Aminopeptidase With a Specific Antibody," Biochem. Biophys. Acta., 389:389, 1975.
48. Louvard, D., *et al.*, "The Brush Border Intestinal Aminopeptidase, a Transmembrane Protein as Probed by Macromolecular Photolabelling," J. Mol.

Biol., 106:1023, 1976.

49. Trier, J. and Madera, J., "Functional Morphology of the Mucosa of the Small Intestine," in Johnson, L., ed., *Physiology of the Gastrointestinal Tract* (New York: Raven Press), V2, 931, 1981.
50. Kenny, A., "Proteinases Associated With Cell Membranes," in Barrett, A., ed., *Proteinases of Mammalian Cells and Tissues* (Amsterdam: North Holland), 393-443, 1977.
51. Kenny, A., "Endopeptidases in the Brush Border of the Kidney Proximal Tubule," in Elliott, K. and O'Connor, M., eds., *Peptide Transport and Hydrolysis* (Amsterdam: Elsevier), 209, 1977.
52. Kim, Y. "Intestinal Mucosal Hydrolysis of Proteins and Peptides," in Elliott, K. and O'Connor, M., eds., *Peptide Transport and Hydrolysis* (Amsterdam: Elsevier), 173, 1977.
53. Rhodes, J., *et al.*, "Intestinal Hydrolysis of Disaccharides and Peptides: Comparison of Hydrolases and Perfusion Studies," in Elliott, K. and O'Connor, M., eds., *Peptide Transport and Hydrolysis* (Amsterdam: Elsevier), 245-259, 1977.
54. Shoaf, C., *et al.*, "Isolation and Characterization of Four Peptide Hydrolysates from the Brush Border of Rat Intestinal Mucosa," Biochem. Biophys. Acta., 445:694, 1976.
55. Kim, Y., "Discussion", in Elliott, K. and O'Connor, M., eds., *Peptide Transport and Hydrolysis* (Amsterdam: Elsevier), 54, 1977.
56. Josefsson, L., *et al.*, "Intracellular Hydrolysis of Peptides," in Elliott, K. and O'Connor, M., eds., *Peptide Transport and Hydrolysis* (Amsterdam: Elsevier), 201, 1977.
57. Matthews, D., "Final Discussion," in Elliott, K. and O'Connor, M., eds., *Peptide Transport and Hydrolysis* (Amsterdam: Elsevier), 353-354, 1977.
58. Ugolev, A., and Gozette, I., XXV International Congress of Physiological Sciences, 9:573 (abstr.), 1971.
59. Schultz, S., and Curran, P., "Coupled Transport of Sodium and Organic Solutes," Physiol. Rev., 50:637, 1970.
60. Munck, B., "Intestinal Transport of Amino Acids," in Kramer, M., and Lauterbach, F., eds., *Intestinal Permeation* (Amsterdam: Excerpta Medica), 123, 1977.
61. Cheeseman, C. and Smyth, D., "Specific Transfer Process for Intestinal Absorption of Peptides," J. Physiol., 229:45, 1973.
62. Matthews, D., "Special Features of Peptide Transport," in Elliott, K. and O'Connor, M., eds., *Peptide Transport and Hydrolysis* (Amsterdam: Elsevier), 10, 1977.
63. Matthews, D. and Adibi, S., "Peptide Absorption," Gastroenterology, 71:151, 1976.
64. Rubino, A., *et al.*, "Intestinal Transport of Amino Acid Residues of Dipeptides. 1. Influx of the Glycine Residue of Glycyl-L-Proline Across Mucosal Border," J. Biol. Chem., 246:3542, 1971.
65. Cheeseman, C. and Parsons, D., "Intestinal Absorption of Peptides. Peptide

Uptake by Small Intestine of *Rana pipiens*," Biochem. Biophys. Acta., 373:523, 1974.

66. Sigrist-Nelson, K., "Dipeptide Transport in Isolated Brush Border Membrane," Biochem. Biophys. Acta., 394:220, 1975.
67. Meister, A., "On the Enzymology of Amino Acid Transport," Science, 180:33, 1973.
68. Tate, S. and Meister, A., "Interaction of Gamma-Glutamyl Transpeptidase With Amino Acids, Dipeptides, and Derivatives and Analogues of Glutathione," J. Biol. Chem., 249:7593, 1974.
69. Meister, A., "Biochemistry of Glutathione," in Greenberg, D., ed., *Metabolism of Sulfur Compounds, Metabolic Pathways* (New York: Academic Press), 101-188, 1975.
70. Meister, A. and Tate, S., "Glutathione and Related γ-Glutamyl Compounds; Biosynthesis and Utilization," Ann. Rev. Biochem., 45:559, 1976.
71. Meister, A., *et al.*, "On the Function of the γ-Glutamyl Cycle in the Transport of Amino Acids and Peptides," in Elliott, K. and O'Connor, M., eds., *Peptide Transport and Hydrolysis* (Amsterdam: Elsevier), 123-143, 1977.
72. Palekar, A., *et al.*, "Decrease in Glutathione Levels of Kidney and Liver After Injection of Methionine Sulfoximine into Rats," Biochem. Biophys. Res. Commun., 62:651, 1975.
73. White, op. cit., 409.
74. Ibid., 633.
75. Cornell, J. and Meister, A., "Glutathione and γ-Glutamyl Cycle Enzyme in Crypt and Villus Tip Cells of Rare Jejunal Mucosa," Proc. Natl. Acad. Sci., 73:420, 1976.
76. Matthews, D. and Burston, D., "Intestinal Transport of Peptides," in Kramer, M. and Lauterbach, F., eds., *Intestinal Permeation* (Amsterdam: Excerpta Medica), 136, 1977.
77. Adibi,, S. *et al.*, "Evidence for 2 Different Modes of Tripeptide Disappearance in Human Intestine: Uptake by Peptide Carrier Systems and Hydrolysis by Peptide Hydrolysis," J. Clin. Invest., 56:1355, 1975.
78. Munro, H. and Crim, M., "The Proteins and Amino Acids," in Goodhart, R. and Shils, M., eds., *Modern Nutrition in Health and Disease* (Philadelphia: Lea and Febiger), 67, 1980.
79. Wacker, H. and Semenza, G., "A Brush Border-Bound Peptidase and Amino Acid Transport," in Elliott, K. and O'Connor, M., eds., *Peptide Transport and Hydrolysis* (Amsterdam: Elsevier), 109, 1977.
80. Josefsson, L., *et al.*, "Intracellular Hydrolysis of Peptides," in Elliott, K. and O'Connor, M., eds., *Peptide Transport and Hydrolysis* (Amsterdam: Elsevier), 199-207, 1977.
81. Maroux, S., *et al.*, "Final Discussion," in Elliott, K. and O'Connor, M., eds., *Peptide Transport and Hydrolysis* (Amsterdam: Elsevier), 361, 1977.
82. Noren, O., "Studies on a Soluble Dipeptidase From Pig Intestinal Mucosa. Structural Properties," Acta. Chem. Scand., B28:787, 1974.
83. Sjostrom, H., "Enzymatic Properties of Pig Intestinal Proline Dipeptidase,"

Acta. Chem. Scand., B28:802, 1974.
84. Adibi, S., "Metabolic Fate of Circulating Dipeptides," in Elliott, K. and O'Connor, M., eds., *Peptide Transport and Hydrolysis* (Amsterdam: Elsevier), 265-285, 1977.
85. Conrad, M., *et al.*, "Polymerization and Intraluminal Factors in the Absorption of Hemoglobin-Iron," J. Lab. & Clin. Med., 68:659, 1966.

CHAPTER 9

EVIDENCE FOR TWO MINERAL ABSORPTION PATHWAYS

IN Chapter 7 support was presented for a secondary pathway in the intestine for the absorption of minerals that are chelated to amino acids—a pathway that is distinct and separate from the well known pathway for the uptake of metal ions. Numerous studies were cited to illustrate these distinctions.

From these studies it is evident that for the ionic pathway to function it requires (1) the ionization of the metal salt; (2) chelation of the cation by the integral protein embedded in and transversing the mucosal cell membrane; (3) some diffusion, but mostly active transport of the ion across this membrane through the use of a mineral pump similar to the sodium pump; (4) entry into the mucosal cell. The integral protein complex is subjected to a pH change with the resulting release of the cation from the peripheral protein and its oxidation; (5) rechelation of the metal ion by storage peptides located in the terminal web; (6) movement of the ion to the basement membrane; (7) another change in the cytoplasmic pH resulting in the metal's release from the storage peptides and its consequential reduction; (8) rechelation of the ion by peripheral proteins located in the basement membrane, and finally; (9) transport of the ion to the plasma.

Figure 9-1 reviews this concept of ionic mineral absorption using iron as the example. The iron is in the ferrous state (+2) in the gut when it is chelated to the integral proteins. As it enters the mucosal cell, it is oxidized to the ferric state (+3) and then chelated with apoferritin. The iron is reduced again to the ferrous state at the basement membrane. As it is passed into the plasma, it is oxidized to the ferric state and chelated with transferrin.[1]

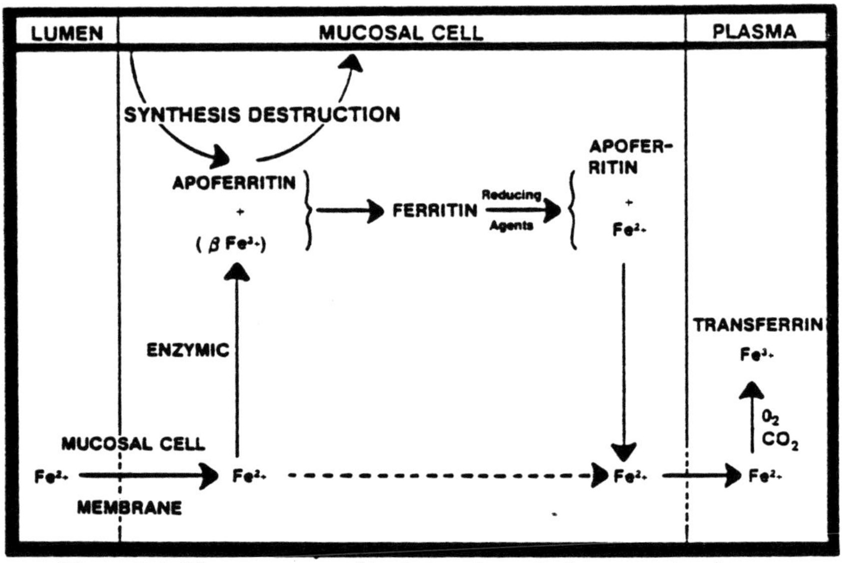

Figure 9-1. The movement of ionic iron from the lumen to the plasma.

The following photographs showing absorption into the mucosal cells suggest that more than one absorption mechanism is active. In the first two figures sections of guinea pig duodenal mucosa obtained 14 days following ingestion of ferrous carbonate are shown.[2] Figure 9-2, a magnification of 7,000 times, shows large particles of the iron in the lumen still not absorbed. The arrows in the cytoplasm refer to ferritin aggregates. In Figure 9-3 these aggregates are magnified 36,000 times. Figure 9-3 is the iron ferritin discussed in Figure 9-1.

Figure 9-4 shows rat duodenum that was excised and exposed to an iron dipeptide-like amino acid chelate for 30 minutes. The dark streaks on the microvilli apear to be the absorbed chelate which is passing from the microvilli into the terminal web of the mucosal cell. This streaking is not observed in duodenum exposed to equivalent amounts of iron as $FeCl_2$ solution for 30 minutes as seen in Figure 9-5. The magnification for Figures 9-4 and 9-5 was 29,250.[27]

EDAX readings, which are X-ray analysis of the mucosal cell

taken from a transmission electron microscope, tend to confirm the visual conclusions relating to Figures 9-4 and 9-5. In the chelate group there was a mucosal cell mean iron content of 62.5 ± 9.1 compared to 48.6 ± 11.5 for the inorganic iron.[27] This is a 28.6% increase.

The above research suggests that more than one absorption pathway for iron is operational in the duodenum. Had there been only one uptake mechanism employed for absorption of both forms of iron, then both the photographs and the EDAX would have been more similar. Their dissimilarity suggests two iron absorption pathways.

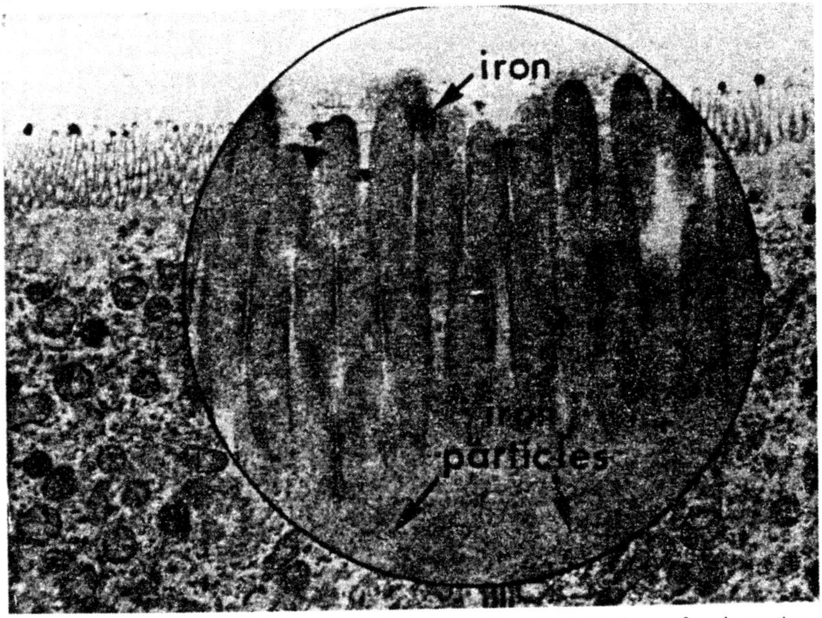

Figure 9-2. Iron carbonate from the lumen unabsorbed 14 days after ingestion. The arrows in the cytoplasm refer to iron ferritin. The magnification is 7,000 times.[2]

The ionic intestinal absorption model does not, however, account for the observed differences between the absorption of minerals as ions and those presented to the mucosa as chelates (with two amino acids per each mineral atom to form a stable dipeptide-like molecule.)

As shown in the *in vitro* jejunal absorption experiment discussed

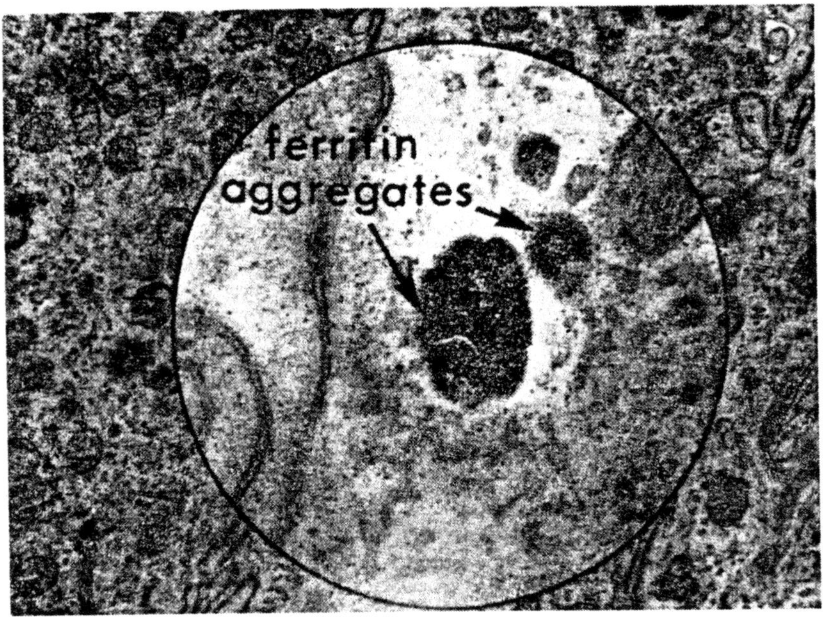

Figure 9-3. Ferritin aggregates in apical cytoplasm (arrow). The magnification is 36,000 times.[2]

in Chapter 7, the mineral in the chelate was absorbed at a more rapid rate than the mineral from the equivalent metallic salts. Consequently, a second model that acts simultaneously with the ionic absorption model is needed to account for the observations made on the absorption of these non-precipitated and yet non-gut ionizable mineral chelates.

Based on what is known of the absorption of dipeptides, the chelate absorption model probably requires (1) chelation of the divalent cation with two amino acids to form a dipeptide-like low molecular weight acid chelate that has been stabilized. (If it is not stabilized, the chelate will break apart in the gut and the released cation will be absorbed, if at all, via the ionic absorption pathway;[3] (2) the attachment of the peptide chelate at the terminal amino site to a specific membrane carrier peptide believed to be glutathione; (3) active transport across the brush border of the mucosal cell; (4) once absorbed into the mucosal cell there can be (a) total or partial hydrolysis of the chelate into amino acids and a cation with the released metal being chelated by an intercellular storage peptide (such as

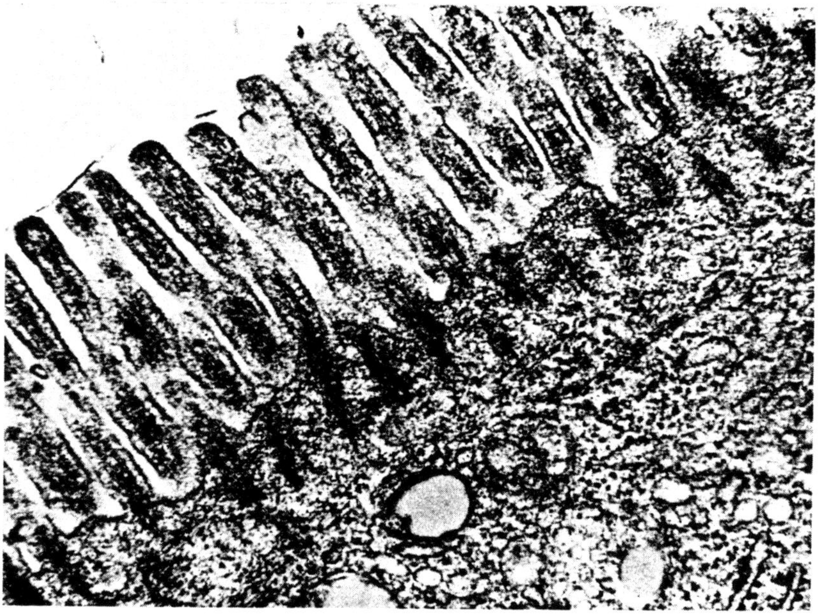

Figure 9-4. The absorption of iron into the microvilli from a dipeptide-like iron chelate. The dark streaks are the absorbed iron. Magnification was 29,250 times.[27]

apoferritin in the case of iron) and transported to the plasma as outlined in the above ionic absorption model, or (b) transport of the intact chelate to the basement membrane without hydrolysis (The degree of hydrolysis is dependent upon the stability constant of the chelate at various pH's in the cytoplasm and to the extent to which the cation in the chelate blocks the cytosol dipeptidase enzymes.); (5) at the basement membrane there can be (a) total or partial hydrolysis of the chelate into amino acids and ion with the chelation of the released ion by the basement membrane integral proteins and transportation to the plasma, or (b) transport across the mucosal cell basement membrane as the original dipeptide-like amino acid chelate after attachment to glutathione in the basement membrane. These possibilities are shown in Figure 9-6.

This and the following chapters are concerned with presenting a series of experiments which verify and elucidate the secondary non-ionic absorption mechanism for chelated minerals. Three basic types of studies will be examined. The first is a series of *in vitro* demonstra-

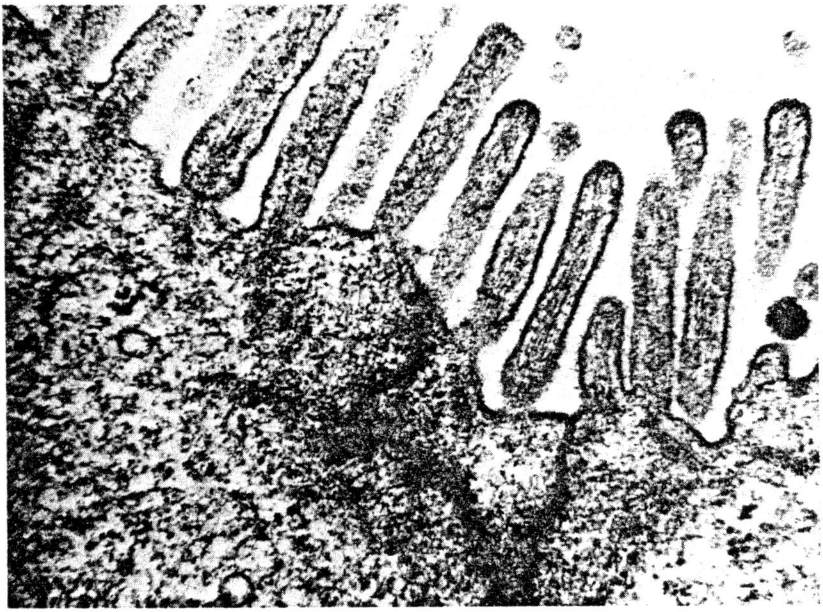

Figure 9-5. The absorption of iron from $FeCl_2$. Notice the absence of the streaking seen in Figure 9-4. Magnification was 29,250 times.[27]

tions of the ability of the mucosal cells to absorb minerals as intact dipeptide-like amino acid chelates with a comparative analysis of the efficiency of that mechanism versus the efficiency of absorbing the same minerals in the ionic state. This study will demonstrate the existence of two absorption mechanisms in the mucosal cells.

The second studies are *in vivo* experiments using radioactive isotopes. Through comparative measurements of isotopes in the blood as a function of time, two general absorption areas in the small intestine that operate independently of each other and are possibly governed by different controls will be examined.

The third group of studies will elucidate the fate of the chelates after mucosal cell absorption. These *in vitro* studies will also use radioactive isotopes as part of the dipeptide-like amino acid chelates. These experiments will further support the concept that the mode of absorption of this specific type of chelate which can be found in food classified them as peptides rather than minerals as far as intestinal absorption is concerned. These three groups of experiments will clearly elucidate the coexisting secondary absorption mechanism in

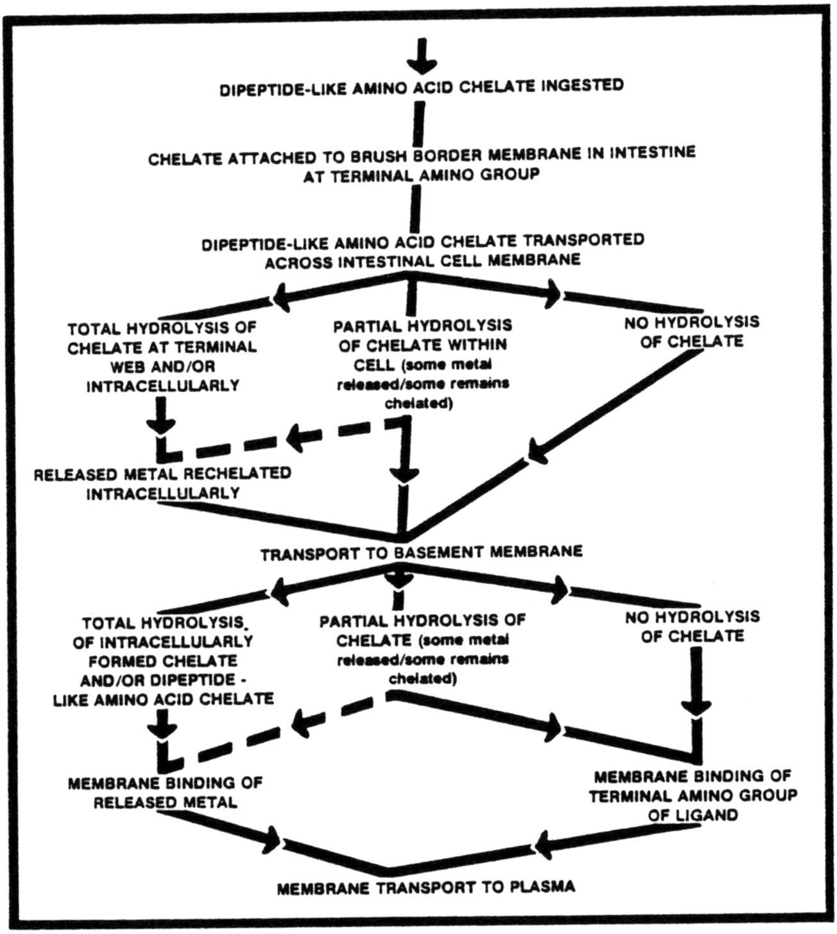

Figure 9-6. Possible fates of an ingested dipeptide-like amino acid chelate.

the mucosa which is independent of the ionic absorption mechanism.

By way of introduction, it should be recognized that there is no general agreement concerning the intestinal absorption mechanism of a chelated mineral. Cardon stated that the role of a ligand was to chelate the cation and maintain it in a sequestered state until its arrival at the ionic absorption site in the duodenum. At that point the ligand would for some unexplained reason release the cation and it would be absorbed through the normal cation absorption sites.[4]

Cardon was not the only researcher to expouse the "protection role" of the chelate to the metal ion. Ashmead, *et al.*[5] published studies in 1974 which not only demonstrated greater absorption of the dipeptide-like chelates, but also explained their superior absorption with a model similar to Figure 9-1. They suggested the role of the ligand was to carry the cation to mucosal cells where the mineral was absorbed by active transport. No hypothesis as to exactly how this active transport process occurred was presented. Nevertheless, the discussion tended to support the Cardon theory that the ligand prevented the cation from entering into unwanted precipitating chemical reactions in the gut. It was assumed that the metal had to retain or at least be able to regain its ionic state before it could be absorbed.

The idea of the chelate utilizing a second pathway was not considered. As a result, the use of chelates for improved mineral absorption has generated controversy.

The formulation of chelates by synthetic ligands such as ethylene-diamine-tetraacetic acid (EDTA), or by amino acids or other organic acids is a well established phenomenon.[6] Indeed, as early as 1893 the technique of bonding metals to ligands by chelation was observed.[7] In 1920 Morgan and Drew named this chemical phenomenon "chelation" derived from the Greek word "chele" meaning claw.[8]

Since these original pioneering studies, chelation has been used to good advantage in several industrial applications. In the 1950s and early 1960s several investigators first attempted to use the mineral sequestering principles of chelation to improve the utilization of minerals in the feed of animals.[9] Some attempts were marginally successful.[10] Others were not.[11]

In each case, the rationale appeared to be one of sequestering the mineral until it arrived at the appropriate sites on the duodenum where it was absorbed via the ionic mineral absorption mechanism. The main question was what caused the ligand to release the metal ion at that particular moment? The active transport mechanism that moves cations across the mucosal cell membrane requires the presence of a free cation in order to operate. If the stability constant of the chelate was too high, the mineral would not be released from the ligand and the transport mechanism could not be activated. Thus, little meaningful absorption would occur.

On the other hand, if the stability constant was too low, dissociation of the metal and the ligand could occur in the stomach. Thus the very chemical reactions between the cation and precipitating anions that the chelate was supposed to prevent would occur. The net result would be less than enhanced mineral absorption.

Stability constants are obviously an important criterion for protection of the chelates. But what is the proper stability constant for the chelate? Furthermore, what mechanism changes the pH so drastically at the luminal side of the mucosal cell membrane at exactly the brush border as to cause the metal to be released only then and not before? And finally, why will this unknown mechanism not affect the pH of the rest of the intestinal environment and cause a premature breaking of the covalent bonds of the chelate? Obviously the difficulty in answering these questions makes the protection-release role of the chelate suspect.

Even if the above questions could be successfully answered, there is still a question of different absorption rates of chelates made under varying conditions. As shown in Chapter 7, 13.2% of a crude iron-porphorin chelate having a molecular weight of about 64,500 was absorbed compared to only 1.7% absorption of an iron-ferritin chelate whose molecular weight was about 500,000.[12] The molecular size of the chelate has a great deal to do with its absorption. Thus, even though the same ligands were used to chelate the minerals (as determined by chemical analysis), the results varied.[5] Before the effects of molecular weight on absorption efficiency was recognized, the absorption variances caused interpretation of the role of chelates to remain unclear.

There was also the question of the ligand used for chelation. Not all ligands have the same physiological effect. For example, in work reported by Martinez-Torres and Layrisse,[13] it was suggested that the amino acid, cysteine, was the most efficient ligand for chelation of iron if improved intestinal absorption of that metal was desired.

White, *et al.*[1] have suggested that there may be a "mucosal block" that regulates the quantity of ionic iron entering the mucosal cell at any given time. While this may be true, could one not postulate a "mucosal block" for certain ligands? The intestinal absorption cell may partially or totally reject a given ligand due to competition for absorption sites.[14] That rejection would also include the rejection of any cation chelated to the rejected ligand.

This is true of amino acids. The work of MacInnis, *et al.*[14] showed that cells have specific absorption sites for many amino acids with some overlap in affinity. The major exception is methionine. It is a strong competitor for, and is absorbed at all amino acid absorption loci.

Not only is there a possible block to certain amino acids, but even the chelation of a cation with one amino acid rather than two may have an effect on the outcome of the metal's absorption. Peptides use an intestinal transport mechanism that is different from, and independent to, the amino acid transport system.[15] A chelate of a metal ion with a single amino acid is a compound different from that formed when the ratio of metal to amino acids is 1:2. It is no longer an amino acid *per se*. Where are they absorbed? Do the amino acid absorption sites reject this chelate as they do peptides? Is it absorbed as a peptide? Is there competition for absorption sites?

While the answers to these questions are not completely clear, it has been demonstrated that the absorption of a metal chelated to two amino acids to form a dipeptide-like molecule is more efficient than the chelation of that same metal to a single amino acid.[3] It is also known that a chelate having a molecular weight of over 1500 (three plus amino acid ligands) is not absorbed intact[16] presumably due to the fact that most tripeptides and larger molecules are digested on the luminal side of the mucosal membrane, at the brush border, prior to absorption.[17]

From the above it is evident that these and other questions tend to emphasize the extreme importance of comprehending how a mineral that has been chelated with amino acids is absorbed in the mucosa. That edification is important to better understand how minerals that are chelated to amino acids in food can be absorbed as well as to improve overall mineral nutrition through proper chelation.

The first experiments designed to answer this question examined intestinal mineral absorption *in vitro*. In Chapter 7 a series of experiments was described in which adult male Sprague-Dawley rats were used. Rather than repeat the experimental design, suffice it to say intestinal segments removed from these rats were incubated *in vitro* in conditions that simulated the body. Absorption of dipeptide-like chelates into these jejunal segments was compared to that of sulfates (SO_4), oxides (O_2), and carbonates (CO_3). The chelates contained $10.9 \pm 0.3\%$ of a specific metal ion of copper, magnesium, iron, or zinc. The inorganic salts contained an equivalent amount of metal. Exposure of the various metals to the intestinal segments was for two minutes.

The results of the type of experiment described in Chapter 7 are shown in Table 9-1 and represent the mean of several experiments. Each value represents mg% of metal per dry weight of tissue. This experiment was repeated several times. In each experiment the results were essentially duplicable.

Table 9-1.

MEAN UPTAKE OF CHELATES COMPARED TO MINERAL SALTS INTO RAT JEJUNAL SEGMENTS IN VITRO

	Chelate	SO_4	O_2	CO_3	Control
	$\bar{x}$	$\bar{x}$	$\bar{x}$	$\bar{x}$	$\bar{x}$
Copper	35	8	11	6	Trace
Magnesium	94	36	23	51	7
Iron	298	78	61	82	53
Zinc	191	84	66	87	14

Restating Table 9-1 in which the mean assay of jejunal segments for mineral absorbed from solutions of each metal salt or the amount present in the control is considered to be 100%, and compared to the mean assay of minerals from the chelates in the jejunal segments, Table 9-2 can be developed. Basically this second table shows how much greater the absorption of the chelate is compared to an equivalent amount of metal salt or to the control which was not exposed to any metal.

From Table 9-1 and 9-2 it is obvious that when the mineral is chelated with two amino acids to form a molecule that is similar to a dipeptide (the metal which occupies a central position links the two amino acids), the resulting jejunal uptake is much greater than when the metal is left in the ionic state.

One of the conclusions that can be drawn from these series of experiments is that the chelate was absorbed by a different pathway than the ion. If both shared the same pathway, then one would anticipate a similar uptake because each jejunal segment was exposed to an equivalent amount of metal for equal time periods.

Admittedly if this rationale is strictly adhered to, even the ionic

Table 9-2.

COMPARATIVE RAT JEJUNAL UPTAKE OF CHELATE VERSUS INORGANIC MINERAL SALTS AND CONTROL

	SO_4	O_2	CO_3	Control
Copper Chelate	4.4 x	3.2 x	5.8 x	35.0 x
Magnesium Chelate	2.6 x	4.1 x	1.8 x	13.4 x
Iron Chelate	3.8 x	4.9 x	3.6 x	5.6 x
Zinc Chelate	2.3 x	2.9 x	2.2 x	13.6 x

forms of the same metal do not comply. There are variations in the uptake of the ions. Nevertheless,, these variations appear to be due more to interference from the anions (SO_4, O_2, CO_3) in the enzymatic KRB Solution than differences in pathway selection. For example, as cations are absorbed in the carbonate solution, there would be greater affinity for H^+ ions and the possible formation of hydroxyl products which may reduce future uptake. Changes in intestinal environment pH will also affect metal uptake.[19] An elevated pH may accelerate oxidation which in turn reduces absorption.[20] Furthermore, as the cations are absorbed into the intestine, it would change the equilibrium between the cations and anions in solution. This may in turn increase the binding propensity of the greater concentration of the anions which in turn would interfere with the binding of the cations by integral proteins on the mucosal membrane.

In spite of these explanations it should be pointed out that in reality there are not wide variations in uptake when Tables 9-1 and 9-2 are closely examined. The major difference in jejunal mineral absorption resulted from exposure to the chelates. This form of mineral demonstrated much greater uptake than any of the salts.

From this series of experiments it is postulated that the chelates were absorbed via a different pathway than the ions. This conclusion is based on two premises.

First, the chelate did not ionize in the thermolysin-KRB Solution. While an equilibrium state for chelates does exist in solution and cations are constantly being released and rechelated, for all practical purposes no ionization took place. The equilibrium constant

$$Ke = \frac{(MC)}{(M^{++})(L)}$$

where M^{++}) represents the molar concentration of the metal ion, L the molar concentration of the ligand and MC the molar concentration of the chelate in the system provided the theoretical stability constant for the chelates used.[21] Based on these calculations, the chelate was stable in the KRB Solution. Consequently, no measurable amount of free ions was available from the chelate to be absorbed by the ionic pathway.

Now if the chelate remains intact in the solution as indicated above and if one insists the metal must be absorbed only through the ionic pathway, then to follow this rationale there must be a hydrolyzation of the chelate prior to the absorption of the cation. In other words, the metal ion must be freed from the ligands if it is to activate the mechanism that activates the pump for active transport of cations. That rationale leads to the second reason for suggesting that the chelate was taken up by a different pathway than the ions.

The hydrolyzation of the chelate must occur at the luminal side of the mucosal cell membrane at the brush border. The digestion of the chelate is brought about by the excretion of enzymes produced within the intestinal cells. These enzymes will digest peptide molecules into their component amino acids by breaking the linkages.[22,23] Nevertheless, there is little or no hydrolyzation of dipeptides, or even tripeptides. They are most generally absorbed into the mucosal cell intact.[24] It is only when a molecule of the tetrapeptide dimension is presented to the brush border that hydrolysis begins.[17,25]

Membrane digestion of the dipeptide-like chelates into free amino acids and metal ions prior to their jejunal absorption apears very unlikely for three reasons: The first is due to the experimental design in general. It must be remembered that the intestines were washed three times in Krebs-Ringer Bicabonate Buffer (pH 7.4) Solution and then pre-incubated for one minute in a similar solution into which oxygen and carbon dioxide were bubbled for equilibration purposes. Presumably any unintegrated external enzymes usually located on the brush border[22] would have been removed by these four separate washings. Thus, with no external means to break the dipeptide-like chelate into its individual components, it would be required to be absorbed intact into the mucosal cells if it was taken up at all.

The second reason is based on observations discussed in Chapter 8. It was noted there that the bonding of a metal to a dipeptide blocks the dipeptidase activity within the mucosal cell. This results in a "smuggling" of the mineral into the cell with the intact dipeptide.

The third reason for the premise that the chelate was absorbed intact can be deduced if one assumes that the metal was released from its two amino acid ligands prior to membrane transport. Then its transport would be carrier dependent in exactly the same manner as the ions from the carbonate, the sulfate, and the oxide.

If that actually occurred—that is a membrane digestion of the chelate and transport of the released ions by active transport through the ionic pathway—then ionic absorption from the metal salts should be greater than the chelate as a function of time because membrane transport of the metal from the chelate would have required a second step, the dissociation of the chelate, before ionic absorption could take place. If that second step had actually occurred, it would have delayed the mucosal uptake of the chelate by the amount of time it took to hydrolyze the chelate and free the metal. Thus as a function of time, ionic absorption should be greater than chelate absorption. But to the contrary, Figures 9-7 to 9-10 show that the absorption of the chelates was far greater than the absorption of the equivalent amounts of ions during the same time period.

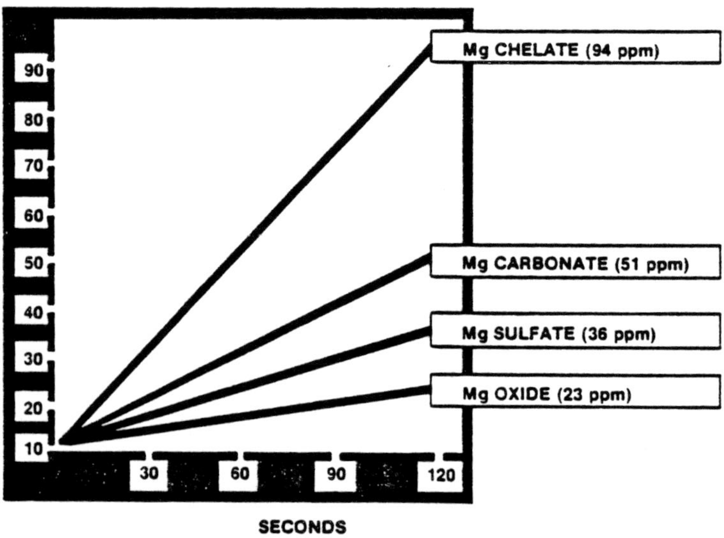

Figure 9-7. Intestinal uptake of magnesium.

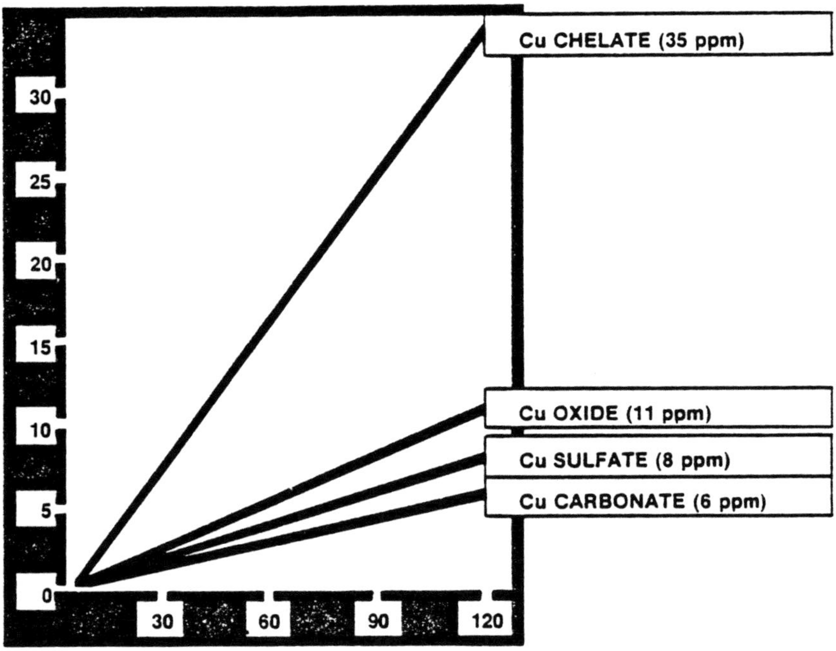

Figure 9-8. Intestinal uptake of copper.

From the above discussion it is evident that the chelate did not dissociate in the KRB Solution or at the mucosal cell membrane. For the minerals to be absorbed as ions into the jejunal sections dissociation must occur. It is evident that there was uptake of the chelates into the jejunum because Table 9-1 shows that there were 35 times as much copper, 13 times as much magnesium, 6 times as much iron, and 14 times as much zinc in the jejunal sections as was found in the control.

The greater amount of metal in the mucosal cells indicates absorption of the chelates, and because the chelates did not dissociate and release the metals prior to absorption, the metals could not have been taken up via the ionic pathway. Thus, one must conclude that the dipeptide-like amino acid chelates were absorbed intact by a different mechanism than used for ions. The existence of at least two absorption pathways for minerals into the mucosa cells is evident. One allows the mineral to be absorbed as part of a dipeptide-like molecule in which it occupies a central position between two amino

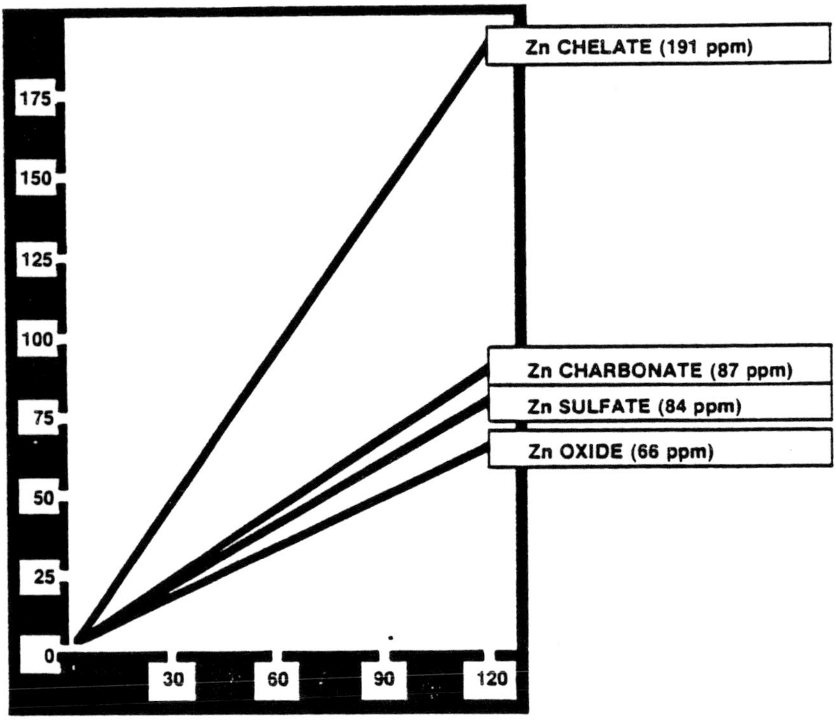

Figure 9-9. Intestinal uptake of zinc.

acids and is held in place by covalent bonds between the amino acids and the ion. The other pathway requires metallic ionization and chelation by integral proteins on the brush border of the mucosal cell membrane prior to mineral absorption.

Since tracer studies indicate there is no competition between the two pathways,[26] and since the chelate is absorbed as an intact molecule which is similar in nature to a dipeptide, one is led to the conclusion that the second pathway for mineral uptake is probably the peptide absorption pathway discussed in Chapter 8.

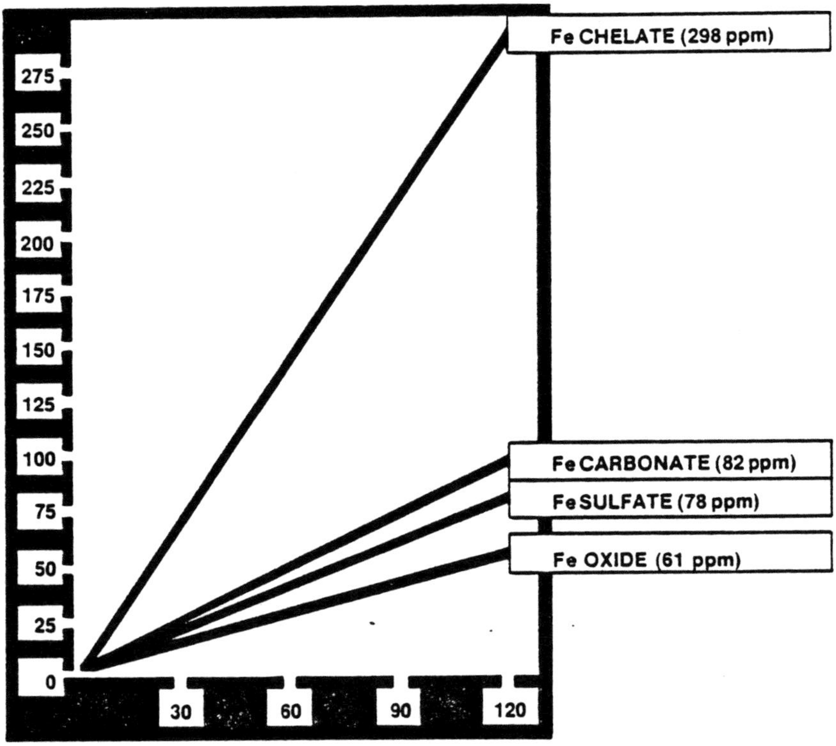

Figure 9-10. Intestinal uptake of iron.

REFERENCES

1. White, A., *et al., Principles of Biochemistry* (New York: McGraw Hill), 866, 1973.
2. Freeman, J. A. and Greer, J. C., "Intestinal Fat and Iron Transport, Goblet Cell Mucus Secretion, and Cellular Changes in Protein Deficiency Observed with the Electron Microscope," Am. J. Digest Dis., 10:1004, 1965.
3. Jensen, N., "Biological Assimilation of Metals," U.S. Patent No. 4,167,564, Washington, D.C., Sept. 11, 1979.
4. Cardon, B., "The Importance of Chelation in Trace Minerals," a paper given at the Montana Nutrition Conference," Feb. 11, 1963.
5. Ashmead, H., *et al.*, "Chelation Does Not Guarantee Mineral Absorption," J. Appl. Nutr., 1974.
6. Mahler, W. and Corbes, F., *Biological Chemistry* (New York: Harper & Row), 196, 1966.
7. Werner, A., Anorg U. Allgem. Chem., 3:267, 1893.

8. Morgan, G. and Drew, H., "Residual Affinity and Coordination. II. Acetylacetones of Selenium and Tellurium," J. Chemical Soc. 117:1456, 1920.
9. Scott, M., "Chelates in Animal Nutrition," paper given at Animal Nutrition Research Council Meeting, Washington, D.C., 1965.
10. Scott, M., "Increasing Utilization of Zinc," Animal Nutrition and Health, 17:17, June 1962.
11. Miller, R., "Chelating Agents in Poultry Nutrition," Paper given at Delmarva Nutrition, Short Course, March 1968.
12. Layrisse, M., *et al.*, "Ferritin Iron Absorption in Man," Blood, 45:689, 1975.
13. Martinez-Torres, C. and Layrisse, M., "Effect of Amino Acids on Iron Absorption From a Staple Vegetable Food," Blood, 35:669, 1970.
14. MacInnis, A., *et al.*, "Specificity of Amino Acid Transport in the Tapeworm Hymenolepis Diminuta and its Rat Host," Rice University Studies, 62:182, 1977.
15. Rubino, A. and Guandalini, S., "Dipeptide Transport in the Intestinal Mucosa of Developing Rabbits," in Elliott, K. and O'Connor, M., eds., *Peptide Transport and Hydrolysis* (Amsterdam: Elsevier), 79-81, 1977.
16. Tiffin, L., "Translocation of Micronutrients in Plants," in Dinauer, R., ed., *Micronutrients in Agriculture* (Madison: Soil Science Society), 207, 1972.
17. Burston, D., *et al.*, "Mechanisms of Peptide Transport," in Elliott, K. and O'Connor, M., eds., *Peptide Transport and Hydrolysis* (Amsterdam: Elsevier), 79-81, 1977.
18. Agar, W., *et al.*, "The Uptake of Amino Acids by the Intestine," Biochemica et Biophysica Acta, 14:80, 1954.
19. Briggs, G. and Calloway, D., *Nutrition and Physical Fitness* (Philadelphia: W.B. Saunders Co.), 252, 1979.
20. Narins, D., "Absorption of Nonheme Iron," in Bezkorovainy, A., *Biochemistry of Nonheme Iron* (New York: Plenum Press), 99, 1980.
21. Kratzer, F., "Chelation-Its Influence on Nutrient Utilization," Feedstuffs, 37:62, Feb. 6, 1965.
22. Ugolev, A., *Physiology and Pathology of Membrane Digestion* (New York: Plenum Press), 1968.
23. Adibi, S. and Kim, Y., "Peptide Absorption and Hydrolysis," in Johnson, L., ed., *Physiology of the Gastrointestinal Tract* (New York: Raven Press), V2, 1078, 1982.
24. Silk, D., "Amino Acid and Peptide Absorption in Man, in Elliott, K. and O'Connor, M., eds., *Peptide Transport and Hydrolysis* (Amsterdam: Elsevier), 16:20, 1977.
25. Williams, S., *Nutrition and Diet Therapy* (St. Louis, C.V. Mosby Co.), 1974, 1977.
26. Fang, S., *et al.*, "Bioavailability of Zinc: Effect of Amino Acid Chelation," paper presented at Albion International Symposium, Salt Lake City, 1980.
27. Graff, D. and Hess, W., "Intestinal Absorption of Iron Studied by Electron Microscope," Unpublished, 1984.

CHAPTER 10

THE INTESTINAL ABSORPTION OF DIPEPTIDE-LIKE CHELATES AS SMALL PEPTIDES

IN Chapter 9 it was shown that membrane transport or mucosal cell uptake of the dipeptide-like amino acid chelated minerals was more rapid than the uptake of similar quantities of metallic ions. This greater rate of absorption was hypothesized to be due to the pre-chelated state of the metals which led to their consequential absorption as peptide-like molecules whereas the metallic ions were first bound to membrane integral proteins before they could be taken into the mucosal cells by a different pathway which is slower. Thus, one step in the uptake of the mineral, that of chelation at the brush border of the mucosal cell membrane, was eliminated.

In previously published studies it has been shown that the absorption of ionic minerals into the mucosal cell is a two-step process.[1-5] The first step, as indicated above, involves the chelating of the ion to the integral proteins transversing the mucosal cell membrane. In this bound state the metal is carried across the membrane into the cell, principally by the process of active transport as discussed in Chapter 6.

Referring to metal ions, Sahagian, *et al.*, have reported, "the transport of a given metal, or its accumulation in the serosal solution, is largely dependent upon the degree of affinity of the metal for binding sites available on the tissue, either on the outer or inner cellular surfaces."[6]

These investigators have reference to both steps in the uptake of metal ions. The binding sites on the outer surface of the cell membrane are the integral proteins. The binding sites on the inner surface of the cell membrane are the storage and carrier proteins, such as apoferritin for iron, that transport the ions from the terminal web to the basement membrane.

Once the ion is on the internal side of the membrane at the terminal web it is released from the transport protein molecule of the membrane and oxidized. The cation is then re-chelated with another intracellular protein that functions for both storage and ultimate movement of the metal to the basement membrane of the cell. This second binding of the metal ion at the terminal web is also the second step in the two-step uptake of cations from the gut.

Using zinc as an example, it was demonstrated in Chapter 9 that not only was uptake of minerals into the cell more rapid in the form of dipeptide-like amino acid chelates compared to ions, but also the transfer across the cell to the plasma was more rapid. The reason for that more rapid transport was hypothesized to be due to a different intestinal absorption site for the chelates as well as a different metabolic treatment of the chelate by the mucosal cells. Because of its molecular configuration, the dipeptide-like amino acid chelate appears to be a "smuggler" complex and is regarded as a dipeptide by the absorptive mechanism of the mucosal cells rather than a metal.

Obviously, the above hypothesis relates only to the uptake of the chelate. Most cells in the body are equipped to hydrolyze dipeptides that are absorbed from the lumen into the plasma intact. Consequently, the chelate can be dissociated in the same manner as the dipeptides in these same cells. Once that occurs, the released cation is probably handled in the very same manner as the body would normally handle any absorbed mineral. It would transport, use, excrete, and store according to its needs.

With that brief review, the purpose of this chapter is to present evidence that demonstrates that the mucosal uptake of the dipeptide-like amino acid chelate is not dependent upon a chelation process at the brush border as are free ions. Furthermore, this dissimilarity is further proof of two pathways for metal uptake.

For this study four adult male Sprague-Dawley rats were used. They were maintained separately in identical wire bottom cages. Each of them weighed 180 grams, plus or minus ten grams.

All of the rats were fed a commecial laboratory rat chow (Allied Mills) *ad libitum*. Their diet was considered to be nutritionally adequate and none evidenced any signs of nutritional deficiencies.

On the day of the experiment the rats were divided into two groups. Each rat was anesthetized with ether and then sacrificed by cervical dislocation. The small intestine of each rat was excised and washed three times in Krebs-Ringer Bicarbonate Buffer (pH 7.4) Solution (KRB).[7] All intestinal contents were removed by these washings. The intestines were then everted to allow them to be exposed to the mucosal solutions in the vessels in which they would ultimately be placed. Using the bath as a mucosal solution facilitated injecting isotopes into the intestinal environment for test purposes. After being everted and prior to being placed in the mucosal bath, they were placed in a well-oxygenated (95% oxygen and 5% carbon dioxide) KRB Solution. Each intestine was cut into five sections of 15 centimeters in length.

The segments were tied at both ends and transferred to vessels containing a buffer (mucosal) solution (pH 7.4) as shown in Figure 10-1.[8] Using this system environmental conditions similar to those normally found in the intestine were duplicated under controlled conditions.

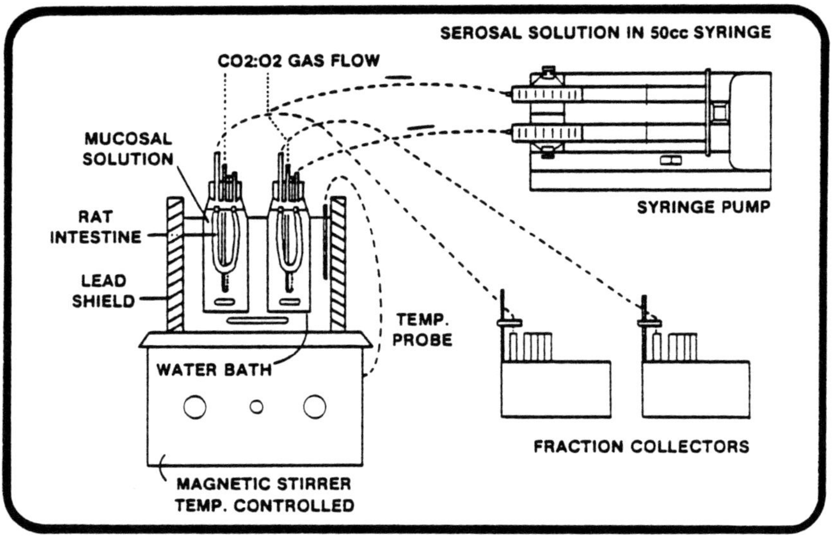

Figure 10-1. Experimental procedure for mineral uptake in intestinal loops.

The water bath surrounding the vessels was maintained at a constant 37°C. An oxygen (95%) and carbon dioxide (5%) gas mixture was bubbled through the vessels at all times. Serosal solution was injected into the inverted intestines by means of automatic 50cc syringes with a flow rate of approximately 50 microliters per minute. Collection of the zinc transferred from the mucosal solution to the serosal solution was accomplished by fraction collectors.

In five replications the intestinal segments were used to measure the mucosal-to-serosal transfer of ionic zinc alone while the other five baths were used to measure the muscoal-to-serosal transfer of zinc chelated to L-histidine to form a dipeptide-like molecule. The control solution consisted of zinc chloride (1×10^{-4}M) containing 50 microliters of ^{65}zinc (0.5 MCi/μl) as a tracer. The experimental vessel contained an equivalent amount of zinc and ^{65}zinc tracer plus histidine (7×10^{-4}M). The molar ratio of zinc to histidine was one mole of metal to two moles of histidine.

Every hour for five hours the amount of ^{65}zinc that was transferred from the mucosal solution in the vessel to the serosal solution in the fraction collectors was assayed by liquid scintillation.

After the data were collected from the first study the experiment was repeated again as outlined above except the oxidative phosphorylation uncoupler, 2, 4-dinitrophenol (DNP) was added to mucosal solutions of both reaction vessels at the concentration of 4×10^{-4}M. DNP will stop the uptake of metal ions by preventing the formation of ATP which supplies the necessary energy for active transport of the metal ion as discussed in Chapter 6[9] and in Chapter 8.[18]

The active transport of a mineral is defined as "an enzyme-catalyzed reaction occurring on one side of the membrane that is linked to the vectoral transport of a substance not chemically altered by the process".[9] Metabolic energy in the form of ATP (adenosine tripolyphosphate) is derived from the oxidative phosphorylation in the mucosal cell mitochondria. This energy is provided to the carrier molecules in the brush border which then are able to move the ion across the mucosal cell membrane through a series of conformational changes.[10,11] In the case of the peptide, the energy is supplied to the compounds that are manufactured after the peptide has been released in the cytoplasm.

At the end of the five-hour collection period, the data were compiled from the serosal solutions and the means determined for the uptake of ionic zinc with and without the DNP and the uptake of the zinc chelate with and without the DNP. The effect of chelating zinc with L-histidine on the rate of transfer from the mucosal solution to the serosal solution is summarized in Figure 10-2.

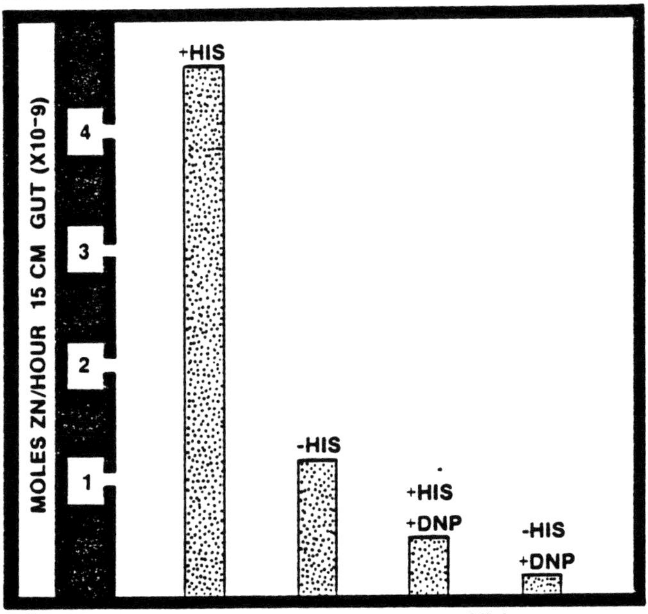

Figure 10-2. Intestinal transport of ^{65}zinc.

When histidine is bound to zinc as a dipeptide-like amino acid chelate, the transfer of the zinc across the intestine increases about four times compared to absorption of the zinc as an ion is in solution alone. This suggests not only a greater uptake of the zinc into the mucosal cells, but it also shows a more rapid transport across those cells to the serosa. This supports the conclusions made concerning the research as discussed in Chapter 9.

If DNP is added, the mucosal-to-serosal transfer of zinc decreases to about 5% of normal. In this case, the addition of increased amounts of histidine to the mucosal solution does not appreciably increase the uptake of the zinc. These findings confirm the reports of other investigators[12,13] and indicates that both the ion and the chelate require ATP for absorption. It is interesting to note that even though both forms of zinc require ATP for their active transports, there appears to be some diffusion into the mucosal cells. Passive diffusion does not require ATP. In the dipeptide-like chelated form the diffusion is greater than in the ionic form suggesting that if there is a regulatory mechanism in the mucosal cell, it is more blocking for the ion than when the metal is "smuggled" in as part of two amino acids.

If the last part of this study is discussed first, it becomes apparent that, as was indicated in earlier chapters, chelation of the cation is an essential requirement for its transfer across the membrane of the mucosal cells. It was stated that ionic zinc as well as other minerals must be bound to the integral proteins in the cell membrane by the process of chelation before they are transported across the membrane to the interior of the cell.

"The transport of a given metal or its accumulation in the serosal solution is largely dependent upon the degree of affinity of the metal for binding sites available on the tissue, either on the outer or inner cellular surfaces, and the total capacity of the tissue for binding or retaining a given metal. Therefore, any factor which inactivates tissue-binding sites or alters the binding characteristics of the metallic ion by virtue of luminal chelation, competition, or some other means results in lower tissue retention of that metal."[14]

The DNP does not interfere with this external chelation process. The ion is bound, but it cannot be transported across the cell membrane to any degree because energy is lacking to do so. Thus, there is little transfer of the metal to the serosa. This is summarized in Figure 10-3.[19]

On the other hand, active transport of the chelate also requires energy from ATP. When DNP blocks its production, absorption of the dipeptide-like amino acid chelate is also affected. The following structure[18,24] reviews the roles of ATP in the uptake of the chelate.

Nevertheless, diffusion of the chelate into the mucosal cells, a process that does not require energy, is greater for the chelate than for the free ions. This suggests that perhaps there is a different pathway for diffused chelates than for diffused ions. More data outside the scope of this work are needed before any concrete conclusions could be drawn on that supposition.

The practical application of these results lie not in the effects of DNP, *per se*, but in the effects of ingestion of other substances that

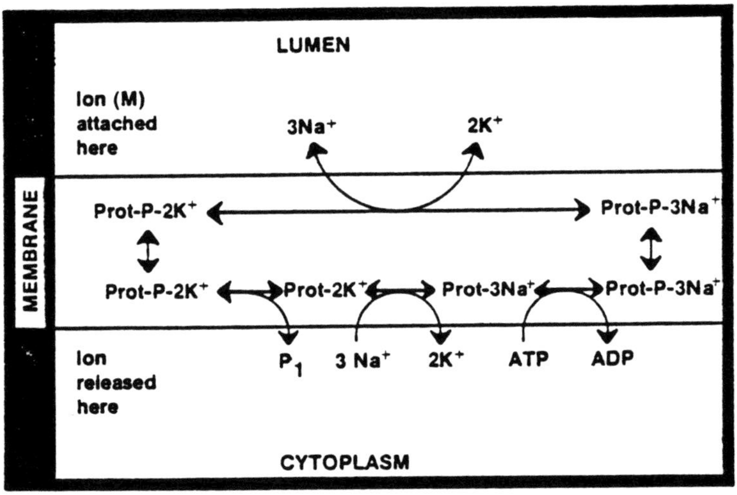

Figure 10-3. The enzymatic reactions involving the carrier protein in the cell membrane.

1. Glutamic acid + cysteine + ATP $\xrightarrow[\text{synthetase}]{\gamma\text{-glutamylcysteine}}$ γ-glutamylcysteine + ADP + P_i

2. γ-Glutamylcysteine + glycine $\xrightarrow[\text{ATP}]{\text{glutathione synthetase}}$ γ-glutamylcysteinylglycine (glutathione) + ADP + P_i

3. Glutathione + dipeptide-like chelate $\xrightarrow[\text{transpeptidase}]{\gamma\text{-glutamyl}}$ γ-glutamyl dipeptide-like chelate + cysteinylglycine

4. γ-Glutamylamino acid $\xrightarrow[\text{cyclotransferase}]{\gamma\text{-glutamyl}}$ 5-oxoproline + dipeptide-like chelate

5. Cysteinylglycine + H_2O $\xrightarrow{\text{peptidase}}$ cysteine + glycine

6. 5-Oxoproline + ATP + $2H_2O$ $\xrightarrow[Mg^{2+}(Mn^{2+}),\ K^+(NH_4^+)]{\text{5-oxoprolinase}}$ glutamic acid + ADP + P_i

precipitate or otherwise complex the zinc so that it cannot ionize and be chelated by the membrane proteins. The net result is the same: lower transfer of the metal from the mucosa to the serosa.

Since zinc deficiencies have been documented in man and animals, the pre-chelating of the zinc (or other cations) with amino acids in a molar ratio of 1:2 (metal to amino acids) would offer an alternative which, in most cases, results in greater absorption because the metal is no longer as free to enter into the precipitating chemical actions in the gut.

To illustrate, Prasad, *et al.* have published detailed clinical studies of eleven Iranian males who were anemic and suffering from zinc deficiencies caused by consumption of a diet consisting of a large amount of unleavened bread. The bread contained phytate which forms insoluble zinc phytate precipitates in the gut that are unabsorbable. (Leavening destroys the phytate.) At the age of twenty their sex organs were still infantile, and they were mentally retarded.[20]

Zinc is required in large amounts by the brain for proper functioning. It is particularly high in the hippocampus.[21] Using histochemical techniques, Ibata and Otsuka found zinc was mainly in the terminal vesicles of the nerve endings of the hippocampal formation.[22] Thus zinc is somehow involved in the proper functioning of the brain, probably in the transmission of nerve impulses. A deficiency results in retardation.

The sex organs of boys are very high in zinc in their mature state. "During puberty, as the prostate, seminal vesicles and testes develop, the body's borderline zinc supply is taxed severely to provide all the zinc needed for male sex glands to function."[23] Hypoganadism can result from a zinc deficiency.

The role of zinc in DNA and RNA and growth have been discussed elsewhere in this book. Suffice it to say that zinc deficiencies result in growth retardation. In fact, in another study of 17 male dwarfs suffering from zinc deficiencies caused by dietary phytates, when nine were given a normal hospital diet plus 100 mg of $ZnSO_4$ daily, they developed normal sexual functions in 59 days compared to 224 days for the other group receiving the hospital diet without the zinc supplement. They also grew 10.5 cm compared to the 4.2 cm for the non-supplemented group during that same period.[20]

Pfeiffer has developed a table of possible clinical disorders resulting from a dietary zinc deficiency.[23] This is seen in Table 10-1.

Table 10-1.

CLINICAL DISORDERS RESULTING FROM A ZINC DEFICIENCY IN MAN

SYNDROME	INVESTIGATOR	PLACE
Poor circulation	J.H. Henzel	U.S.A.
Childhood hyperactivity	Allan Cott	U.S.A.
Epilepsy	A. Barbeau	Canada
Diabetes	E.J. Underwood	Australia
Enlarged prostate	I.M. Bush et al.	U.S.A.
High cholesterol	H.G. Petering	U.S.A.
Effects of oral contraceptives	R. Alfin-Slater	England
Psoriasis	Voorhees et al.	U.S.A.
Hypogonadism	Prasad	Iran
Hypogonadism	Caggiano et al.	U.S.A.
Wound healing	W. Pories	U.S.A.
Loss of taste and smell	R.I. Henkin	U.S.A.
Cadmium poisoning	H.A. Schroeder	U.S.A.
Cirrhosis of liver	Sullivan et al.	U.S.A.
Parenteral nutrition	M.E. Shils	U.S.A.
Dwarfism	A. Prasad	Iran
Poor growth	Hambidge	U.S.A.
Cutaneous striae	Pfeiffer et al.	U.S.A.
Pregnancy, nausea	Pfeiffer et al.	U.S.A.
Kwashiorkor	S. Kumar and K.S. Jaya-Rao	India
Acanthosis	Hallbrook	Sweden
Pyroluria, white spots in nails, lack of dream recall, disperceptions, hallucinations, or depression	Pfeiffer et al.	U.S.A.
Retinal detachment	Pfeiffer (postulated)	U.S.A.
Reye's syndrome	Pfeiffer (postulated)	U.S.A.
Acrodermatitis enteropathica	E.J. Moynahan	England
Acne	Pfeiffer et al.	U.S.A.
Sickle cell disease	S. Dash et al.	U.S.A.
Lack of ovulation	Pfeiffer et al.	U.S.A.
Impotency & amenorrhea	Pfeiffer et al.	U.S.A.
Coarse hairs-eyebrows	F.M. Dementzis	Greece
Body and breath odor	F.M. Dementzis	Greece

Not only does the chelate help avoid dietary caused zinc complexes from resulting in zinc deficiencies, but it is absorbed faster. The rate of transfer of the metal from the mucosa to the serosa is more rapid because the two-step absorption mechanism generally recognized as essential for ionic mineral uptake and transport[1] appears partially or totally avoided. No chelation at the brush border of the cell membrane is required so membrane transport is more rapid. There appears to be little or no release of the metal from the original chelating ligands once the membrane has been crossed. Presumablly this may be the reason the zinc is transported more rapidly to the serosa after mucosal cell absorption. Using the peptide absorption pathway of the gut rather than the recognized ionic absorption pathway, the two-step ionic absorption mechanism and the need for the critical environment this pathway requires are avoided. This avoidance results in both faster and increased total absorption of the zinc and presumably the other minerals such as iron,[15] manganese,[16] and calcium[17] whose absorptive mechanisms are similar to zinc. These data suggest an entirely different pathway for a metal that is pre-chelated with two or more amino acids forming a dipeptide-like amino acid chelate. The metal appears to be absorbed as a component of a small peptide-like molecule rather than an ion.

REFERENCES

1. Evans, G., *et al.*, "A Proposed Mechanism for Zinc Absorption in the Rat," Am. J. Physiol., 228:501, 1975.
2. Crampton, R., *et al.*, "Observations on the Mechanism of Absorption of Copper by the Small Intestine," J. Physiol., 178:111, 1965.
3. DeGrazia, J., "The Intestinal Absorption of Calcium," in Skoryna, S. and Waldron-Edward, D., eds., *Intestinal Absorption of Metal Ions, Trace Elements and Radionuclides* (Oxford: Pergamon Press), 152, 1971.
4. Conrad, M. and Crosby, W. "Intestinal Mucosal Mechanisms Controlling Iron Absorption," Blood, 22:406, 1963.
5. Sahagian, B., *et al.*, "Effects of Inhibitors and Accelerators on Intestinal Absorption of Divalent Trace Elements," in Skoryna, S. and Waldron-Edward, D., eds., *Intestinal Absorption of Metal Ions, Trace Elements and Radionuclides* (Oxford: Pergamon Press), 333, 1971.
6. Ibid., 335.
7. Agar, W., *et al.*, "The Uptake of Amino Acids by the Intestine," Biochemica et Biophysica Acta, 14:80, 1954.
8. Fang, S., *et al.*, "Bioavailability of Zinc — Effect of Amino Acid Chelation," in Ashmead, D., ed., *Chelated Mineral Nutrition in Plants, Animals and Man* (Springfield: Charles C Thomas), 137, 1982.
9. Hobbs, A. and Albers, R., "The Structure of Proteins Involved in Active Membrane Transport," Ann. Rev. Biophys. Bioeng., 9:259, 1980.

10. Robertis, E., *et al., Cell Biology* (Philadelphia: W.B. Saunders Co.), 474, 1975.
11. Giese, A., *Cell Physiology* (Philadelphia: Saunders College), 407, 1975.
12. Hutagalung, R., *et al.*, "Intestinal Uptake and Transfer of Zinc After 2, 4-Ionitrophenol and Ovabain Supplementation," Arch. Tierernaehr, 27:347, 1977.
13. Schwartz, F. and Kerchgessner, M., "Effect of Organic Ligands of Zinc Absorption *in vitro*," Z. Tierphysiol Tierernaehr Futtermittelkd, 35:257, 1975.
14. Sahagin, B., *et al.*, "Effects of Inhibitors and Accelerators on Intestinal Absorption of Divalent Trace Elements," in Skoryna, S. and Waldron-Edward, D., eds., *Intestinal Absorption of Metal Ions, Trace Elements and Radionuclides* (Oxford: Pergamon Press), 335, 1971.
15. Graff, D., *et al.*, "Mucosal Uptake and Transport of Iron Chelated to Methionine and Lysine," (Unpublished), 1981.
16. Ashmead, H., *et al.*, "Stability Constants of Manganese[54] Chelates and Their Effects on Manganese Metabolism," (Unpublished), 1974.
17. Ashmead, H., *et al.*, "The Influence of Stability Constants on Calcium Metabolism *in vivo*," (Unpublished), 1974.
18. Meister, A., *et al.*, "On the Function of the γ-Glutamyl Cycle in the Transport of Amino Acids and Peptides," in Elliott, K. and O'Connor, M., eds., *Peptide Transport and Hydrolysis* (Amsterdam: Elsevier), 123-143, 1977.
19. White, A., *et al., Principles of Biochemistry* (New York: McGraw Hill Book Co.), 298, 1973.
20. Pfeiffer, C., *Mental and Elemental Nutrients* (New Canaan: Keats Publishing Co.), 220, 1975.
21. Harrison, W., *et al.*, "Trace Elements in Human Brain: Copper, Zinc, Iron, and Magnesium," Clinea Chimica ACTH, 21:55, 1968.
22. Ibata, Y. and Otsuka, N., "Electron Microscopic Demonstration of Zinc in Hippocampus Formation Using Timm's Sulfide-Silver Technique," J. Histochem. Cytochem., 17:171, 1969.
23. Pfeiffer, op. cit., 229.
24. White, op. cit., 633.

CHAPTER 11

DETERMINATION OF THE LOCUS FOR INTESTINAL ABSORPTION OF A DIPEPTIDE-LIKE CHELATE

IN previous chapters it was shown that zinc as well as other micro and macro minerals could be absorbed into mucosal cells through at least two pathways. The pathway which has been extensively studied and reviewed in the literature has been termed the "ionic" pathway. There are several essential steps involved when this pathway is used to absorb ions.

The first step is, of course, the ionization of the mineral. In the *in vitro* experiments discussed in Chapter 9 the solubilizing of the mineral was done in the laboratory. This step actually takes place in the stomach in the presence of hydrochloric acid.[1]

As the cation enters the intestine it is essential that it remain soluble as it comes in contact with the integral proteins that are loosely bound within the membrane.[2] This is seen in Figure 11-1. It is here that the metal ion is chelated to the membrane-bound protein.[3,4]

This chelate is not a molecule that is similar to the dipeptide-like amino acid chelate. First, the molecular weight of the ligand forming the integral protein is probably much higher than the ligands which make up the dipeptide-like amino acid chelate. Using calcium as an example, the molecular weight of the calcium-binding protein has been determined to be between 24,000 and 28,000.[5] At the other end of the spectrum the zinc-binding protein in the mucosal cell membrane which has been identified as a complex peptide, has a molecular weight of approximately 1,500.[6] Since the identification of the

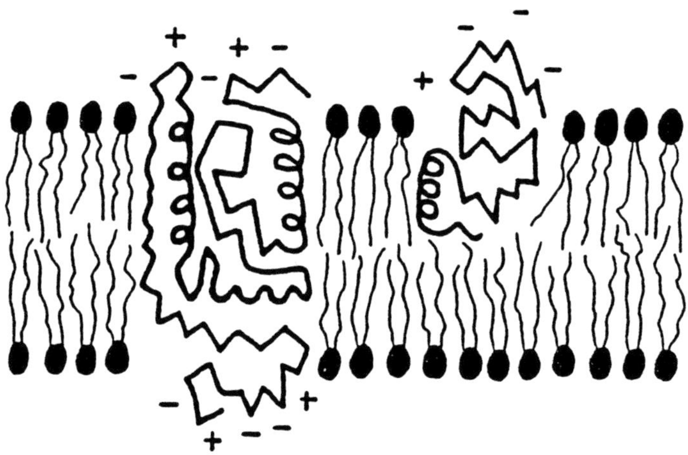

Figure 11-1. The integral protein in the mucosal cell membrane.

zinc-binding protein was reported most recently (1978) and made use of the most sophisticated equipment, it is probable that ultimately the same, or a similar protein, will be identified as that carrier in the membrane necessary for transport of all the cations.

The fact is, however, that this protein has a molecular weight of approximately 1,500. That makes it a ligand that is too large to actively function in the same manner as the dipeptide-like amino acid chelate. For the latter to be absorbed intact into the absorptive cell, it must have a total molecular weight, including the cation, of less than 1,500[7]; that is, the molecule must not be much larger than a dipeptide.[41,42] The integral protein by itself, excluding the metal ion, has a molecular weight of approximately 1,500. When the metal is added and the molecular weight of the protein doubled so that there is a 2:1 ratio of protein molecules to metal atoms, the total molecular weight is in excess of 3,000. That is too large for the integral protein to function as a dipeptide-like molecule in the absorption of minerals, even if it could be released from the membrane.

The second dissimilarity between the integral protein and the individual amino acids in the dipeptide-like chelate is in the chelation itself, as alluded to above. For example, it has been determined that the calcium-binding protein in the mucosal cell has the capacity to bind one mole of calcium per mole of protein.[8] Thus, the chelation of the cation by the integral protein would form a chelate similar to

Figure 11-2,[9] except the protein would be much more complex than seen in this Figure.

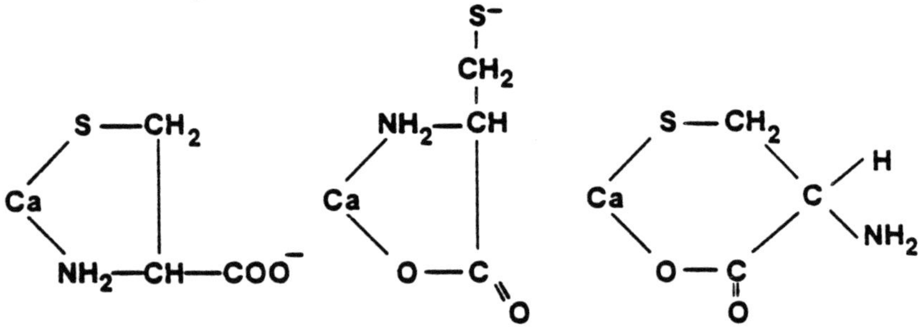

Figure 11-2. The chelation of calcium to form a single heterocyclic ring.

This particular arrangement is suggested because the binding protein in the mucosal cell membrane is composed of several amino acids, namely aspartic acid, glutamic acid, glycine, cysteine and lysine. Of these amino acids it has been postulated that the sulfhydryl group of the cysteine is the most probable position to bind the metal ion present to the mucosal cell from the lumen.[6] Thus, as shown in Figure 11-2, a stable five-member chelate ring involving sulfur is formed between the cation and the integral protein.

In Figure 11-3 an X-ray diffraction reveals the crystalline structure of the dipeptide-like amino acid chelate.[10] There are several differences that are discernable at a quick glance.

Figure 11-3. X-ray diffraction of a dipeptide-like amino acid chelate using glycine as the ligands.

The most obvious is the position of the metal ion. Instead of occupying a terminal position on the peptide, the cation is centralized between two amino acids. The resulting molecule contains a total of ten members forming two rings rather than the single five-member ring in the terminal case of Figure 11-2. This results in two major differences between the chelates. First, removing the metal ion from a terminal position in the peptide and placing it within the peptide increases the stability of the chelate. The greater the number of rings formed with a given metal, the greater the stability of the chelate.[11] The second difference brought about by the central position of the ion is the difference in the molecular configuration of the resulting chelated mineral molecule. Where the ion is between two amino acids, the resulting molecule tends to behave as a dipeptide. The metal loses all of its metallic properties. On the other hand, when the metal ion is at the terminal position, it is still electronically active. Other chemical reactions can occur with it. Some metallic properties still remain.

Having examined some of the major differences in the chelates formed by the integral protein and the two amino acids, one is now prepared to examine the differences in uptake of the same metal via these two types of chelates. In the case of the ion, most of the metal is transported across the membrane of the mucosal cell by an active transport process requiring energy.[12] The cell membrane functions as a transducer of chemical energy for the selective transport of the metal ion.[13]

Using zinc as a representative example it is noted that the absorption of ionic zinc into the mucosal cell, like the other minerals, is a two-step process.[3] The first step is for the ion to be transported across the membrane. When it arrives in the cytoplasm, at the terminal web, it is released and re-chelated with another protein molecule.[14] This protein stores and transports the zinc to the basement membrane of the mucosal cell.

The dipeptide-like chelate is absorbed through a second pathway which is independent of the ionic pathway by attaching a terminal amino group (NH_2) of the chelate [1] to a membrane-bound tripeptide, glutathione, at the gamma-glutamyl linkage.[15,16] Six enzymatic reactions involving glutathione, ATP, and other cofactors bind and release the dipeptide-like amino acid chelate at the terminal amino group point as the intact molecule which acts like a dipeptide is quickly moved across the mucosal membrane into the cell in large quantities.

After absorption into the mucosal cell interior there may or may not be a second step in the absorption. If enzymatic hydrolysis of the peptide-like chelate takes place as it frequently does with absorbed peptides,[18] then the zinc-ligand bond may be broken and the ionic zinc picked up by a storage and transport protein similar in function to the intracellular protein described in the second step of the ionic absorption pathway.

It is unlikely that there is complete intracellular dissociation of this particular type of chelate after absorption. As stated earlier, the more rings formed with a single metal ion, the greater its stability.[11] Thus, a dipeptide-like amino acid chelate would be relatively unaffected by the slight decreases in pH found in the mucosal cell that effects the release of the metal ion carried into the cell by an integral protein.[19]

Secondly, the presence of a metal ion such as magnesium, cobalt, manganese, or zinc within a peptide molecule blocks the hydrolysis of the peptide.[20,21,24] To illustrate, a 10% calcium-peptide chelate was prepared[22] and found to be enzyme resistant.[23]

Finally, dipeptides in particular are frequently transported intact to the portal blood from the gut. All cells of the body have dipeptidase activity.[24] In fact, there is evidence to suggest that dipeptides are transported normally from one part of the body to another.[24] Thus, there is no reason to insist that a dipeptide-like amino acid with its high stability is dissociated in the mucosal cell. In fact, there is some evidence to the contrary.[25]

While the experiment in Chapter 9 did not elucidate the exact uptake mechanism for either the ionic or chelated form of mineral, it is believed that the *in vitro* absorption demonstrated by the experiments in that chapter corresponded to the two measured uptake pathways described above. The purpose of the experiments reported here was to gain a clearer understanding of the intestinal processing of the dipeptide-like amino acid chelates. Are they absorbed in the same mucosal cell sites as the ions or at the peptide loci which are located predominantly beyond the ionic absorption sites in the duodenum?[26,27]

To answer these questions sixteen healthy adult male Sprague-Dawley rats each weighing 170 grams, plus or minus 10 grams, were maintained in identical wire screen bottomed cages for several weeks. They were fed a rat chow (Allied Mills Labolox) *ad libitum* until 24 hours prior to the commencement of the experiment. (The rats had also been grown on Labolox.) All were considered normal and had received a nutritionally adequate diet up to the time of the experiment.

Mineral analysis of the Labolox revealed 40.8 mg of zinc per pound, 7.56 mg of copper per pound, and 61.9 mg of iron per pound of feed as determined by atomic absorption spectrophotometry. Based on that analysis, no rat suffered from a zinc deficiency so intestinal absorption of the experimental zinc isotopes to be used in this experiment would occur at a "normal" rate.

If a nutritional deficiency of zinc did occur or if there was an excess of a zinc blocker, such as copper, in the diet so that an artificial zinc deficiency did occur, then the uptake of the zinc isotopes would have been biased by enhancing absorption.

Commencing 24 hours prior to dosing with the zinc, the rats were randomly divided into two groups of eight rats each and fasted to reduce and/or eliminate the formation of insoluble zinc precipitates in the gut which would tend to interfere with zinc absorption. It will be remembered from Chapter 2 that dietary substances such as phosphates,[28] phytates,[29] or other minerals themselves[30] can interfere with intestinal absorption of zinc.

Following the 24-hour fast, each rat in the first group received a solution containing 5.0 micrograms of zinc as zinc chloride of which 20 microcuries of ^{65}zinc was added as a tracer. The eight rats in the other group each received a solution of 5.0 micrograms of zinc as a dipeptide-like amino acid chelate (similar in molecular structure to Figure 11-3) involving two molecules of amino acid chelated to one atom of zinc to form two five-member rings. Like the zinc chloride, each chelate dose contained 20 microcuries of ^{65}zinc. Each rat was given a single dose of zinc (as a chloride or chelate) after anesthetization with nembutal. After laparotomy, zinc isotopes were injected into the duodenum.

Commencing 15 minutes following the dosing, serial blood samples were taken by suborbital bleeding in which a capillary tube was inserted posteriorly and downwards into the orbit of the eye. This allowed many samples of blood to be taken with a minimum of discomfort to the rat. The second bleeding occurred 30 minutes after dosing (8:30); the third took place 45 minutes after dosing (8:45); the fourth at one hour after dosing (9:00); the fifth at two hours after dosing (10:00); the sixth bleeding took place two and one-half hours after dosing (10:30); the seventh bleeding was done three hours after dosing (11:00); and the eighth at four hours after dosing (12:00). The experiment was terminated after the eighth bleeding because previous studies have indicated that zinc uptake reaches a plateau at about five hours.[31] There were 256 samples of blood taken from the rats during the course of the experiment.

At each bleeding 100 microliters of blood were removed and pipetted onto stainless steel planchets labeled as to the time of sampling and from which group it was removed—the chloride or chelate group. The samples were taken to a constant dryness and weight by drying 12 hours in an oven at 100°C. Assay for isotope levels of zinc in the blood of each sample was accomplished with a Nuclear Chicago Model 8731 gamma ray spectrophotometer with a two-inch NaI (TL) scintillation detector and a strip chart recorder. Radiation from each blood sample was measured for five minutes on each sample.

The results of gamma ray spectrophotometry are shown in Table 11-1. The data are presented as corrected counts of gamma radiation per minute and represent the mean of eight animals.

Table 11-1.

MEAN INTESTINAL ABSORPTION OF ZINC CHLORIDE VERSUS ZINC DIPEPTIDE-LIKE AMINO ACID CHELATE AS MEASURED IN BLOOD OF RATS

Time after Dosing Corrected Gamma Counts Per Minute	^{65}Zinc Chloride	^{65}Zinc Dipeptide-Like Chelate
15 Minutes	7	5
30 Minutes	19	13
45 Minutes	16	14
1 Hour	10	14
2 Hours	16	17
2.5 Hours	21	34
3 Hours	22	34
4 Hours	26	36

The results in Table 11-1 were converted to a graph by plotting the corrected counts of gamma radiation per minute on the y axis and time on the x axis. This is seen in Figure 11-4. The curves that were developed from the mean data were interpreted as follows and are based on our understanding of the two different absorptive mechanisms discussed.

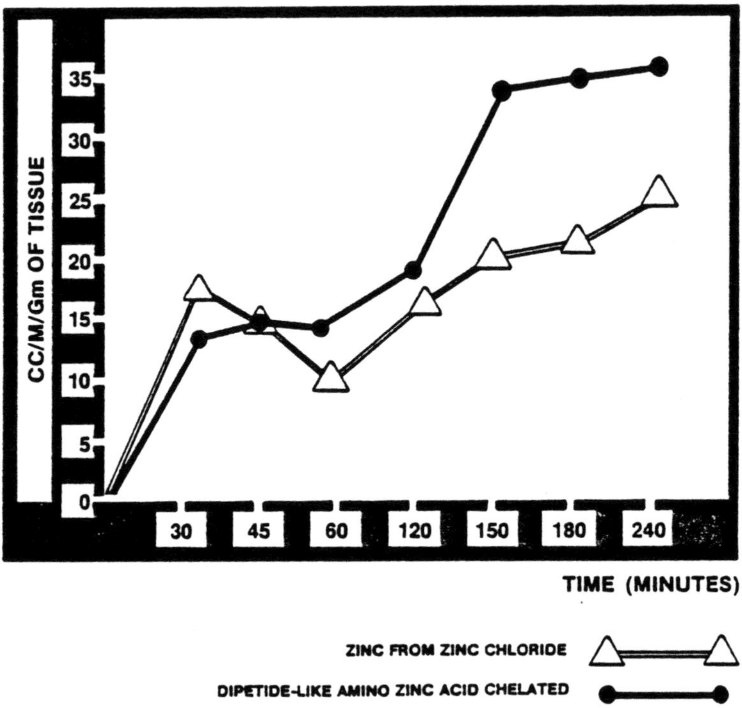

Figure 11-4. Intestinal absorption of zinc (*in vivo*) from ZnCl and zinc dipeptide-like amino acid chelate.)

As a function of time, absorption of the zinc from zinc chloride was initially more rapid. It was observed in the blood in higher quantities earlier than the zinc from the dipeptide-like amino acid chelate. However, there was a decrease inthe zinc blood levels in the zinc chloride group commencing between 25 and 30 minutes after dosing. The zinc blood levels of that group began to rise again at about 40 minutes after dosing and peaked at about two hours and 45 minutes.

While initially slower than duodenal absorption of zinc from zinc chloride treated animals, the zinc levels from the zinc dipeptide-like amino acid chelate treated animals increased throughout the duration of the sampling period and reached a plateau commencing at about 25 minutes (briefly) and after about three hours (another major increase).

The curves in both groups have a basic similarity. Early in their absorption rates the slope both rises sharply and then plateau. At about one hour after dosing, the slope of both curves begin to increase again. At certain points during these increases, both curves have the same slopes although the heights of the slopes are different.

The total uptake of the zinc from the intestines into the blood was 18% higher for the chelate group compared to the chloride group.

Because of the initial higher levels of zinc from the chloride group compared to the chelate group, coupled with the overall higher uptake from the chelate, a delay of some sort was indicated in the absorption of the chelate. That delay was believed to be physiological and not metabolic.

When minerals are in the soluble ionic state, they are absorbed in the upper duodenum of the intestine just below the pyloric valve where the intestinal pH remains low. A low pH is essential to maintain the ionic state of the metal ions. As the minerals descend into the jejunum, the pH rises and solubility is more difficult to achieve. Most of the absorption of protein in the form of amino acids and small peptides occurs further on in the small intestine after the introduction of the pancreatic enzymes and bile which raise the pH. Protein absorption is usually completed in the upper one-third to one-half of the small intestine.[27]

This observation suggests that absorption of the chelate was retarded in the beginning because it had to move through the intestine past the ionic absorption sites to the peptide absorption sites. While it is probable that each mucosal cell is capable of some ionic and/or peptide absorption, the number of specific sites are highly regionalized. Furthermore, luminal environmental factors, such as pH, also influence ionic and peptide uptakes. Thus, with the introduction of both the zinc chloride and the zinc dipeptide-like amino acid chelate into the upper duodenum, it appeared that the zinc chloride was given an unfair advantage as a function of time. The chelate had to descend the intestine before significant absorption could occur while the chloride was absorbed immediately.

In the experiments recounted in the preceding chapter, it was shown that zinc can be absorbed into the intestinal mucosa as an intact dipeptide-like amino acid chelate based on the different rates of uptake. The results of this experiment support that conclusion. If the initial delay in the appearance of blood levels of ^{65}zinc from the chelate were due to dissociation of the chelated zinc into the ionic state in the lumen, then its absorption levels and pattern, including the sharp inclines and declines, should have been identical to the zinc chloride group and retarded only by the time required for ionization and re-chelation by integral proteins. The only difference deductively should have been the delay involving the luminal destruction of the chelate because in this supposed case, the mucosa would still have been exposed to equal quantities of ionic zinc.

It is recognized that due to intestinal motility and changes in pH, this rationale may not be accurate. Nevertheless, in the chelate group there was no sharp decrease in assimilation levels after the initial increase as there was in the chloride group. That drop in the chloride group may have been due to saturation of the mucosal cells with zinc ions from the chloride followed by exchange diffusion as seen in the case of iron loading in Figure 5-6. If that is what happened, it is interesting to note the same reaction did not occur with the chelate. If saturation occurred in the group treated with the chelate, the zinc was not returned to the gut. Thus, it appears that the reason for the lower initial absorption of the chelate can only be explained by a different physiological mechanism which is the longer time it normally takes to reach the peptide absorption sites.

Whereas peptic digestion in the stomach produces mostly peptides from protein,[32] the majority of the peptide bonds are broken after the proteolytic enzymes of the pancreas are secreted into the intestine.[33] This occurs near the bile duct.[27] The pancreatic buffers raise the pH of the duodenum from about 2.0 (which keeps ionic zinc soluble) to a neutral or alkaline pH.[34] When this occurs, zinc absorption from free ionic zinc is reduced[35] because of precipitation of the zinc ions. If the zinc had been released from the chelated state to the ionic state, it would have occurred in the alkaline pH. In that environment there would have been immediate precipitation and correspondingly lower absorption.

This study clearly showed that contrary to what would have been expected if the dipeptide-like amino acid chelated zinc would have been broken apart in the lumen by enzymes or acid, the zinc blood levels from this group were 18% higher than the zinc chloride group. Consequently, it must be concluded that the zinc from the chelate group was not in the free ionic state when it was absorbed and was probably absorbed as an intact chelate in much the same manner as described in the experiments in Chapter 9.

The zinc from the chelate group did not appear in the serial blood samples as rapidly as the zinc from the zinc chloride group. This suggests that the chelated zinc was absorbed in the small intestine at the peptide absorption sites rather than at the ionic absorption sites which are more anterial.

One might argue that because portal blood was not used in this study, no conclusions concerning absorption sites can be drawn. Technically, that is true. But as stated earlier, a close examination of Figure 11-4 reveals a marked similarity between the zinc chloride and the zinc chelate curves. This suggests that while there may have been tissue uptake of the zinc from portal blood (which was confirmed at the conclusion of the experiment by assaying tissues for zinc absorption), there nevertheless was a similar regulatory mechanism at work that allowed a certain percentage of the zinc to be passed to the serial blood. The fact that more of the zinc from the zinc chloride entered the serial blood samples sooner suggests that more also entered the portal blood sooner. Because the mucosal uptake of the chelate is greater than the ionic zinc as a function of time,[36,37] the logical conclusion, especially with reduced intestinal motility from nembutal, is that the absorption site of the chelate is further down the intestine past the ionic zinc site.

While this series of experiments did not definitely answer the question about absorption by the liver, the data strongly suggested the interpretation given above.

It seems reasonable to conclude that the chelated zinc was absorbed through a different locus than the ionic zinc as previously discussed. If the chelated zinc remained bound to the two amino acid ligands (fasting removed other substances), there would have been no attraction for it at the ionic absorption sites. Indeed, in this bound state the zinc is simply the bonding mechanism for two amino acids which form a dipeptide-like compound. On a theoretical basis, mucosal membrane transport of the dipeptide-like amino acid chelate should in no way affect the individual bonds between the amino acids and the metal since membrane transport of this dipeptide-like amino acid chelate is dependent upon attaching it to the carrier molecule, glutathione, at its terminal amino (NH_2) group.[38]

The experiments in Chapter 9 demonstrated the more rapid uptake of zinc from the dipeptide-like amino acid chelate form than from the free ionic state. A close examination of Figure 11-4 suggests a similar phenomenon. During the first 35 minutes after dosing, the zinc blood levels of the zinc chloride group were higher probably because that zinc reached its ionic absorption sites first because they are more anterial and were being absorbed and assimilated.

During that same period, a smaller amount of zinc from the chelate group was being absorbed suggesting a slightly longer period for the majority of that form of zinc to reach its absorption loci further down the intestine. Approximately 35 minutes after dosing, enough of the chelated zinc had passed by the majority of the ionic absorption sites to reach the majority of the peptide absorption loci which are more posterial. It would have been at this point the zinc blood levels of the two groups reversed in concentration due to the more rapid uptake and probably more rapid transport of the zinc in the chelate form (which is supported by results cited in a later chapter). This *in vivo* study suggests that the dipeptide-like amino acid chelates are predominantly absorbed at the peptide absorption sites of the small intestine and not at the ionic absorption loci. Calculations of the rates of motility of the small intestine place the chelate absorption loci at about the same location as the absorption sites of amino acids and peptides.

There is one last point that should be considered. Because there are numerous similarities in the curves of the zinc blood levels from the zinc chloride and the zinc chelate groups in Figure 11-4, it suggests that although not as prominent in the case of the chelate, perhaps a similar zinc-regulating mechanism is at work in the mucosal cells regardless of whether the uptake into those cells is from ionic zinc or chelated zinc. The regulatory mechanism for the chelate would probably become active only as the chelate was dissociated into amino acids within the mucosal cells[39] and the zinc released. At this state it would be picked up by an intracellular protein molecule which would first store and ultimately transport the zinc to the basement membrane.[14,40]

It should also be pointed out, however, that because the curve in the case of the chelate group does not fall in the first hour as did the chloride group, in all probability there is some transport of chelates to the blood as intact molecules which did not undergo dissociation in the mucosal cells probably due to the blocking effect of the zinc.[43,44] When tissue levels were measured for ^{65}zinc at the conclusion of the experiment, the increased tissue levels from the chelate group were higher than the aggregate 18% elevation in the blood. The increased zinc in the tissues of the chelate group during the four hours of the experiment could suggest less digestion of the chelate in the mucosal cells.

In this case, one would predict that the first part of the curve would flatten with increased transport and the second part of the curve to increase because the ionic zinc regulatory mechanism in the mucosal cells was circumvented. This is shown in Figure 11-4. Nevertheless, more studies are required before a more definitive explanation is found.

REFERENCES

1. Huebers, H. and Rummel, W., "Iron Binding Proteins: Mediators in Iron Absorption," in Kramer, M. and Lauterbach, F., eds., *Intestinal Permeation* (Amsterdam: Excerpta Medica), 378, 1977.
2. Singer, S. and Nicolsen, G., "The Fluid Mosaic Model of the Structure of Cell Membranes," Science, 175:720, 1972.
3. Evans, G., *et al.*, "A Proposed Mechanism for Zinc Absorption in the Rat," Am. J. Physiol., 228:501, 1975.
4. Sahagian, B., *et al.*, "Effects of Inhibitors and Accelerators on Intestinal Absorption of Divalent Trace Metals," in Skoryna, S. and Waldron-Edward, D., eds., *Intestinal Absorption of Metal Ions, Trace Elements and Radionuclides* (Oxford: Pergamon Press), 335, 1971.
5. Ebel, J., *et al.*, "Vitamin Mucosa," Am. J. Clin. Nutr., 22:431, 1969.
6. Schricker, B., "Studies of an Endogenous Zinc Chelator of Rat Intestine," Ph.D. Thesis, Univ. of Illinois, 1978.
7. Tiffin, L., "Translocation of Micronutrients in Plants," in Dinauer, R., ed., *Micronutrients in Agriculture* (Madison: Soil Science Society), 207, 1972.
8. Wasserman, R., *et al.*, "Vitamin D-Dependent Calcium-Binding Protein: Purification and Some Properties," J. Biol. Chem., 243:3978, 1968.
9. Martell, A., "The Relationship of Chemical Structure to Metal Binding Action," in Seven, M., ed., *Metal Binding in Medicine* (Philadelphia: J.B.

Lippincott Co.), 17, 1960.
10. Dalley, K. and Johnson, B., "X-Ray Diffraction of Chelates," Unpublished, 1979.
11. Martell, op. cit., 2.
12. Solomons, N., "Zinc and Copper in the Gastrointestinal System," in Karcioglu, Z. and Sarper, R., eds., *Zinc and Copper in Medicine* (Springfield: Charles C Thomas), 277, 1980.
13. Pollock, E., "Fine Structural Analysis of Animal Cell Surfaces: Membrane and Cell Surface Topography," Am. Zool., 18:25, 1978.
14. Freeman, J., *et al.*, "Fine Structure of the Intestinal Epithelial Cell in Iron Absorption," in Skoryna, S. and Waldron-Edward, D., eds., *Intestinal Absorption of Metal Ions, Trace Elements and Radionuclides* (Oxford: Pergamon Press), 396, 1971.
15. Meister, A., "On the Enzymology of Amino Acid Transport," Science, 180:33, 1973.
16. Tate, S., and Meister, A., "Interaction of Gamma-Glutamyl Transpeptidase With Amino Acids, Dipeptides, and Derivatives and Analogues of Glutathione," J. Biol. Chemic., 294:7593, 1974.
17. Schwartz, A., *et al.*, "The Na^+, K^+-ATPase Membrane Transport System: Importance in Cellular Function," in Bronner, F. and Kleinzeller, A., eds., *Current Topics in Membranes and Transport* (New York: Academic Press), V3, 2, 1972.
18. Crim, M. and Munro, H., "Protein," in Hegsled, D., *et al.*, eds., *Present Knowledge in Nutrition* (New York: The Nutrition Foundation), 44, 1976.
19. White, A., *et al.*, *Principles of Biochemistry* (New York: McGraw Hill), 866, 1973.
20. Kim, Y., "Intestinal Mucosal Hydrolysis of Proteins and Peptides," in Elliott, K. and O'Connor, M., eds., *Peptide Transport and Hydrolysis* (Amsterdam: Elsevier), 173, 1977.
21. Joseffson, L., *et al.*, "Intracellular Hydrolysis of Peptides," in Elliott, K. and O'Connor, M., eds., *Peptide Transport and Hydrolysis* (Amsterdam: Elsevier), 201, 1977.
22. Mellander, O., "The Nutritional Significance of Some Peptides," Voeding, 16:336, 1955.
23. Mellander, O. and Folsch, G., "Enzyme Resistant and Metal Binding of Phosphorylated Peptides," in Begwood, E., ed., *International Encyclopedia of Food and Nutrition* (Oxford: Pergamon Press), VII, 569, 1972.
24. Abidi, S., "Metabolic Fate of Circulating Dipeptides," in Elliott, K. and O'Connor, M., eds., *Peptide Transport and Hydrolysis* (Amsterdam: Elsevier), 265-285, 1977.
25. Munck, B., "Intestinal Transport of Amino Acids," in Kramer, M. and Lauterbach, F., eds., *Intestinal Permeation* (Amsterdam: Excerpta Medica), 123, 1977.
26. Cheeseman, C. and Smyth, D., "Specific Transfer Process for Intestinal Absorption of Peptides," J. Physiol., 229:45, 1973.
27. Code, C. ed., *Handbook of Physiology* (Baltimore: Williams & Wilkins Co.),

V3, 1125, 1968.

28. Thompson, D., *Trace Elements in Animal Nutrition* (Skokie: IMC Corp.), 26, 1970.
29. Oberleas, D., *et al.*, "The Availability of Zinc from Foodstuffs," in Prasad, A., ed., *Zinc Metabolism* (Springfield: Charles C Thomas), 225, 1966.
30. Suttle, N., "Trace Element Interactions in Animals," in Nicholas, D. and Egan, A., eds., *Trace Elements in Soil-Plant-Animal Systems* (New York: Academic Press, Inc.), 271, 1975.
31. Antonson, D., *et al.*, "Determination of the Site of Zinc Absorption in Rat Small Intestine," J. Nutr., 109:142, 1979.
32. Code, op. cit., V5, 346.
33. Hendrix, T., "The Absorptive Function of the Alimentary Canal," in Mountcastle, V., ed., *Medical Physiology* (St. Louis: C.V. Mosby Co.) V2, 1162, 1974.
34. Texter, E., *et al.*, "Vascular Factors in Metal Ion Absorption," in Skoryna, S. and Waldron-Edward, D., eds., *Intestinal Absorption of Metal Ions, Trace Elements and Radionuclides* (Oxford: Pergamon Press), 381, 1971.
35. Oelshlegel, F. and Brewer, G., "Absorption of Pharmacological Doses of Zinc," in Brewer, G. and Prasad, A., eds., *Zinc Metabolism* (New York: Alan R. Liss), 310, 1977.
36. Graff, D., *et al.*, "Absorption of Minerals Compared with Chelates Made From Various Protein Sources into Rat Jejunal Slices *in vitro*," Address to Utah Academy Arts, Letters and Sciences, Salt Lake City, April 1970.
37. Ashmead, H., "Tissue Transport of Organic Trace Minerals," J. Appl. Nutr., 22:42, 1970.
38. White, op. cit., 112.
39. Matthews, D., "Rates of Peptide Uptake by Small Intestine," Symposium in Peptide Transport in Bacteria and Mammalian Gut held at Ciba Foundation, London, November, 1972.
40. Suso, F. and Edwards, H., "Binding Capacity of Intestinal Mucosa and Blood Plasma for Zinc," Proc. Soc. Exp. Biol. Med., 137:306, 1971.
41. Burston, D., *et al.*, "Mechanisms of Peptide Transport," in Elliott, K. and O'Connor, M., eds., *Peptide Transport and Hydrolysis* (Amsterdam: Elsevier), 79-81, 1977.
42. Peters, T., "The Subcellular Location of Intestinal Peptide Hydrolysis," in Matthews, D. and Payne, J., eds., *Peptide Transport in Protein Nutrition* (Amsterdam: North Holland), 243-267, 1975.
43. Kim, Y., "Intestinal Mucosal Hydrolysis of Proteins and Peptides," in Elliott, K. and O'Connor, M., eds., *Peptide Transport and Hydrolysis* (Amsterdam: Elsevier), 173, 1977.
44. Kim, Y., "Discussion," in Elliott, K. and O'Connor, M., eds., *Peptide Transport and Hydrolysis* (Amsterdam: Elsevier), 54, 1977.

CHAPTER 12

MUCOSAL CELL TRANSPORT OF DIPEPTIDE-LIKE CHELATES AS PEPTIDES

THE preceding evidence that has been presented has shown that both the uptake of the mineral from the lumen into the mucosal cells and its transport across the cells to the plasma or serosa is enhanced significantly when the mineral is chelated and presented as part of a dipeptide-like molecule. In the case of iron, its uptake under the conditions described is up to 4.9 times greater than absorption of the equivalent of iron as salts. Copper had a rate of absorption that is up to 580% better than equivalent quantities of metal as the salt. Magnesium is absorbed up to 4.1 times greater than magnesium as a salt. All of these were determined by *in vitro* studies.[1,2]

Using isotopes it was determined that mucosal absorption of zinc is up to 3.5 times better in the form of the dipeptide-like amino acid chelate compared to the ion.[3] Calcium absorption was 421% better.[4] And finally, manganese uptake was 3.4 times better as the chelate.[5]

Based on the time studies described in Chapters 9 and 10, it was shown that the improved absorption of the dipeptide-like amino acid chelates is due to an intestinal treatment of the chelate as a peptide rather than an ion. The consequence of this type of uptake is the possible avoidance of the two-step absorption mechanism which is generally recognized as essential for ionic absorption; that is, membrane chelation with an integral protein and transport of the ion as part of the chelate into the cell and secondly the release of the ion and rechelation with an intracellular protein for transport to the basement membrane.[6-10]

Membrane transport of the ion requires the functioning of a pump similar to the sodium-potassium pump. The ion is probably important for the functioning of the pump.

The transfer of the metal ion as part of the dipeptide-like chelate molecule is believed to require a series of enzymatic reactions involving a tripeptide called glutathione known to be critical for absorption of dipeptides.[11] Glutathione is a sulfhydryl molecule which is manufactured within the cell.[12] The terminal amino group of the chelate is attached to the gamma-glutamyl linkage of this tripeptide. The resulting molecule is then acted upon by γ-glutamyl transpeptidase and degraded into γ-glutamyldipeptide chelate and cysteinylglycine. The γ-glutamyldipeptide chelate is then acted upon by γ-glutamyl cyclotransferase which degrades the chelate from 5-oxoproline.[13,14] The chelate is thought then to be inside the cell having been transported through the membrane during these enzymatic processes.

The two enzymes involved in this process are very important. For example, "the gamma-glutamyl transpeptidase may function in the translocation [of the chelate]".[15]

If the dipeptide-like amino acid chelate has affinities like a peptide during uptake and absorption as suggested above, is its absorption mechanism the same as a peptide's in relationship to the amino acids? What is its rate of absorption as compared to amino acids? (Figure 12-1[16]) Is there competition for absorption sites between the individual amino acids and dipeptide-like chelates containing those same amino acids?

In Chapter 8 it was shown that the pathway for peptide absorption can be utilized by amino acids,[17] but that generally they have different noncompetitive pathways.[18] It has also been shown that the uptake of the peptide is more rapid than the uptake of the free amino acids that compose the peptide, although transport from the mucosal cells to the plasma is slower due to some cellular hydrolysis.[19] On the basis of these observations, if the dipeptide-like amino acid chelate behaved as a true peptide, one would predict that its uptake and transport would not only be more rapid for the metal as part of this molecule as has been previously seen, but also its cytoplasmic uptake should be more rapid than free amino acids. The following study was designed to examine this point.

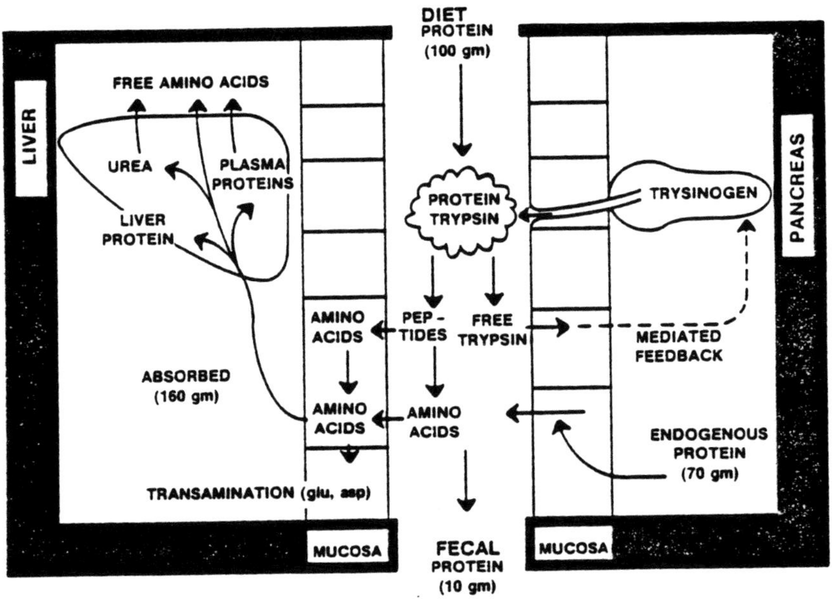

Figure 12-1. Intestinal absorption of peptides and amino acids.

Ten adult male Sprague-Dawley rats which had been maintained in wire bottom cages were randomly selected for the experiment. All had been fed Allied Mills' commercial laboratory rat chow *ad libitum*. All were of equal weight, plus or minus 10 grams.

On the day of experimentation all of the rats were sacrificed by cervical dislocation after first being anesthetized by an intraperitoneal injection of pentobarbital (35 mg/Kg).

Midline incisions were made, and the intestines of each rat exposed. A 10 centimeter section of the ileum was defined. It extended from 15 centimeters to 5 centimeters proximal to the ileocecal valve. This section was excised and removed from each rat.

These excised sections of ileum were thoroughly washed in Krebs-Ringer Buffered Solution (pH 7.4) (KRB) three times to remove all luminal contents. Each was then placed in vessels containing KRB Solution. An O_2-CO_2 gas was bubbled (95% oxygen and 5% carbon dioxide) through the solutions in the vessels. The intestinal segments were everted and tied at both ends with 000 silk thread. They were then quickly transferred to vessels containing buffered mucosal solution as shown in Figure 10-1. This is the same experimental design described in Chapter 10.

The water bath and vessels were maintained at 37°C. Oxygen (95%) and carbon dioxide (5%) were bubbled through the vessels at all times. Serosal solution was injected into the everted intestines by means of 50cc automatic syringes. Collection of the transported zinc which was absorbed from the media with zinc in the form of either zinc chelated to amino acids in the ratio of one atom of zinc to two molecules of amino acids to form dipeptide-like amino acid chelates or ionic zinc and the free amino acid histidine was accomplished with fraction collectors.

The perfusion medium used in this double-labeling study contained ^{65}zinc and ^{14}C-histidine. The zinc concentration on the mucosal side for both groups was 5×10^{-4}M. The concentration of histidine was 7×10^{-4}M. In that ratio of zinc ions to histidine there was 1.06×10^{-4}M zinc to 2.29×10^{-4} histidine.[20]

Serosal samples were removed every hour for five hours and assayed for zinc, histidine, and zinc-histidine complexes by scintillation counting to determine their rates and quantities of transport from the mucosa to the serosa. At the conclusion of the five hours, the intestinal segments were digested and also assayed for mucosal uptake of zinc, histidine and the zinc chelate.

The mean mucosal uptake of the free amino acids of histidine, free zinc ions and the zinc histidine chelate at the end of the five hours is shown in Table 12-1.

The mucosal uptake of the free amino acids of histidine was much less than the uptake of the histidine atoms in the chelate molecule. There was 2.89 times as much histidine from the chelate as there was from the free amino acids. This suggests that the ileum absorbed the chelate as a peptide molecule, and the uptake was more rapid than that of the free amino acids.

The mean amount of ^{65}zinc absorbed in the mucosal cells was significantly greater from the ionic zinc compared to that from the chelate. One might conclude that the absorption of ionic zinc was more efficient than that of the chelate, but when one considers the transfer across the mucosa to the serosa as seen in Table 12-2, more zinc from the chelate was transported than from the ionic form. This suggests that the chelate may be avoiding the two-step process necessary for the absorption and transport of ionic zinc. It also seems to confirm the premise that there is some blockage of dipeptidase activity

Table 12-1.

MUCOSAL UPTAKE OF ZINC, HISTIDINE AND ZINC-HISTIDINE CHELATE		
	Mole/gm Tissue • 5 hour	
Incubation Medium	**Zinc**	**Histidine**
Histidine (7×10^{-4}M)		2.48×10^{-7}
Zinc (1×10^{-4}M)	5.24×10^{-6}	
Zinc + Histidine Chelate	1.88×10^{-6}	7.16×10^{-7}

Table 12-2.

MUCOSAL-TO-SEROSAL TRANSPORT OF ZINC, HISTIDINE AND ZINC-HISTIDINE CHELATE		
	Mole/hr • 15 cm Gut	
Incubation Medium	$\bar{x}$ **Zinc**	$\bar{x}$ **Histidine**
Histidine (7×10^{-4}M)		5.55×10^{-7}
Zinc (1×10^{-4}M)	5.27×10^{-9}	
Zinc + Histidine Chelate	1.4×10^{-8}	2.03×10^{-7}

in regards to the chelate due to the presence of the zinc in the chelate with the two amino acids.

As can be seen from Table 12-2 when studied in conjunction with Table 12-1, the mean mucosal uptake of free ^{14}C-histidine molecules was much slower than the absorbance of the ^{14}C-histidine in the form of a dipeptide-like chelate. Nevertheless, as would be predicted if one was dealing with true peptides, the transfer of the dipeptide-like chelate to the serosal solution was slower than the transfer of free histidine.

When two molecules of histidine are bound in a double heterocyclic chelate ring, the resulting compound appears to act in a manner that is similar to a small dipeptide as far as mucosal uptake and absorption are concerned. In the chelated form the histidine uptake increases but transfer decreases.

Nevertheless, there is probably some intracellular hydrolysis of the dipeptide-like amino acid chelate. The hydrolysis of the chelate that has occurred may be due to the high increase in histidine concentration in the serosal solution compared to its concentration in the mucosal solution. The mucosal ratio of zinc to histidine was 1:2.16. In the serosal solution it was 1:1.5. As Matthews[21] has indicated, many dipeptides that are absorbed intact into the mucosal cells are broken into amino acids after uptake. This appears to be partially true for dipeptide-like amino acid chelates as well, based on the data presented in Tables 12-1 and 12-2. The hydrolysis probably occurred either as the chelate was cleaved from the glutathione carrier as γ-glutamyl cyclotransferase hydrolyzes the γ-glutamyl-dipeptide-like chelate upon entering the cell or at the basement membrane by a peptidase enzyme prior to the chelate's exit from the mucosal cell. In either case, with the cleavage came a slowdown in the transport to the serosa of the resulting amino acids from the chelate when compared to free amino acid transport.

Although there are no data to conclusively prove the exact location of the dipeptide-like amino acid chelate hydrolysis, there is an indication that much of it may take place at the basement membrane of the mucosal cell. This conclusion is based on the amount of zinc within the mucosal cells and the amount of zinc transported to the serosal fluid.

The ratio of zinc in the mucosal cells was 1:2.79 (ionic ^{65}zinc to chelate ^{65}zinc). In the serosal solution the ratio of ionic ^{65}zinc to the chelate ^{65}zinc was 1:0.38. This is 7.41 times more zinc from the chelate in the serosal solution than in the mucosal cells. If the zinc had been released from the dipeptide-like amino acid chelate immediately after entering the cell at the terminal web, the resulting ionic zinc would have been bound to an intracellular protein molecule. Its transport to the basement membrane would have been accomplished by the same carrier proteins that are normally used to move ionic zinc.[6] If that occurred, one could predict that the ratio of zinc for the chelate to the ion within the mucosal cells to the serosal solution would have remained somewhat constant because in that case the intracellular transport mechanism would have been the same. This is particularly important when one considers that there was 7.90 times less histidine from the dipeptide-like amino acid chelate in the serosal solution as a ratio of free amino acid histidine to histidine in the chelate as in the mucosal cells. The disparity in the ratio suggested a lack of dipeptidase activity at the terminal web and the use of a different carrier mechanism for the zinc within the mucosal cells. Furthermore, due to the rapidity at which the chelate appeared in the serosal solution suggested fewer carriers of the chelated zinc to the serosal solution. Thus, the logical choice for the predominant carrier was the original ligand—the histidine amino acids which are bound to the zinc.

According to this study, much of the zinc was not being transported across the basement membrane as the original dipeptide-like chelate, although some of it may have been since small molecules can be transported into the serosa intact.[11] Since there was not an equivalent zinc to histidine ratio in the serosal fluid, as there was in the mucosal cells, the dipeptide-like chelate was clearly not being transported from the intracellular brush border membrane to the basement membrane by a specific intracellular zinc collector and transport protein. Some of it was transported as the original chelate without undergoing any intracellular hydrolysis.

Thus, the zinc must be absorbed into the cell as an intact chelate and transported to the basement membrane as part of this same molecule. If hydrolysis occurred at all, the zinc present at the basement membrane of the cell was released from the amino acid ligands and re-chelated with a transport peptide molecule in the basement membrane. It could then be transported across that membrane into the serosa.

REFERENCES

1. Graff, D., *et al.*, "Absorption of Minerals Compared with Chelates Made From Various Protein Sources into Rat Jejunal Slices *in vitro*," Paper presented at Utah Academy of Arts, Letters, and Science, Salt Lake City, April 1970.

2. Ashmead, H., "Tissue Transportation of Organic Trace Minerals," J. Appl. Nutri., 22:42, 1970.
3. Fang, S., *et al.*, "Bioavailability of Zinc: Effect of Amino Acid Chelation," in Ashmead, D., ed., *Chelated Mineral Nutrition in Plants, Animals and Man* (Springfield: Charles C Thomas), 137, 1982.
4. Ashmead, H., *et al.*, "The Influence of Stability Constants on Calcium Metabolism *in vivo*," (Unpublished), 1974.
5. Ashmead, H., *et al.*, "Stability Constants of Manganese^{54} Chelates and Their Effects in Manganese Metabolism," (Unpublished), 1974.
6. Evans, G., *et al.*, "A Proposed Mechanism for Zinc Absorption in the Rat," Am. J. Physiol., 228:501, 1975.
7. Granick, S., "Iron Transport Mechanism in the Intestinal Tract," Ross Pediatric Res. Conf., October 20-21, 1965.
8. Hutaglung, R., *et al.*, "Intestinal Uptake and Transfer of Zinc After 2, 4-donitrophenol and Oybain Supplementation," Arch. Tierernaehr, 27:347, 1977.
9. Underwood, E., *Trace Elements in Human and Animal Nutrition* (New York: Academic Press), 73, 1977.
10. Ogata, E., *et al.*, "Basis of Action of Hormones on Calcium Absorption," in Skoryna, S. and Waldron-Edward, D., eds., *Intestinal Absorption of Metal Ions, Trace Elements and Radionuclides* (Oxford: Pergamon Press), 359, 1971.
11. White, A., *et al.*, *Principles of Biochemistry* (New York: McGraw Hill), 112, 1973.
12. Ibid., 409.
13. Meister, A., "On the Enzymology of Amino Acid Transport," Science, 180:33, 1973.
14. Tate, S. and Meister, A., "Interaction of Gamma-Glutamyl Transpeptidase with Amino Acids, Dipeptides, and Derivatives and Analogues of Glutatione," J. Biol. Chem., 249:7593, 1974.
15. White, op. cit., 633.
16. Nasset, K., *et al.*, "The Digestion of Proteins *in vivo*," J. Nutr., 56:83, 1955.
17. Munro, H. and Crim, M., "The Proteins and Amino Acids," in Goodhart, R. and Shils, M., eds., *Modern Nutrition in Health and Disease* (Philadelphia: Lea and Febiger), 67, 1980.
18. Ugolev, A. and Gozette, I., XXV International Congress of Physiological Sciences,9:573, (Abstr.), 1971.
19. Robinson, J. and Alvardo, F., "Comparative Aspects of the Interactions Between Sugar and Amino Acid Transport Systems," in Kramer, M. and Lauterbach, F., eds., *Intestinal Permeation* (Amsterdam: Excerpta Medica), 156, 1977.
20. Hallman, P., *et al.*, "The Computer Distribution of Copper (II) and Zinc (II) Ions Among Seventeen Amino Acids Present in Human Blood," Biochem. J., 121:549, 1971.
21. Matthews, D., "Rates of Peptide Uptake by Small Intestine," Symposium in Peptide Transport in Bacteria and Mammalian Gut held at Ciba Foundation, London, November 1972.

CHAPTER 13

INTRACELLULAR HYDROLYSIS OF THE DIPEPTIDE-LIKE CHELATE

IN Chapters 9 through 11 it was shown that as part of a dipeptide-like amino acid chelate, a metal ion was absorbed into the mucosal cell by a pathway which was different from the one described in the literatue for free cations. In Chapter 12 it was shown that as part of this dipeptide-like molecule, zinc was absorbed into the mucosal cell without the molecule undergoing any dipeptidase hydrolysis. The data also suggested that these same ligands that are bound to the metal ion to convey it into the mucosal cell, continue to be bound to it and also transport it to the basement membrane of the cell. It was there that through enzymatic dipeptidase chelate hydrolysis the chelated metal was first released, if it was not transported to the plasma as an intact chelate.

The question remains, however, is chelate hydrolysis in the mucosal cell essential to the serosal or plasma absorption of the metal, or is it a consequence of the amino acids selected as ligands for the chelate? One must remember from Chapter 4 that as ligands change, so does chelate stability.[1,2,10]

Within the epithelial cells of the small intestine are found several peptidase enzymes that are capable of breaking the final peptide linkages of absorbed peptides prior to entering the portal blood.[3] The enzymatic hydrolysis of a peptide bond generally involves the breaking of a covalent bond.[4] The ionic bond, if one exists, is broken by changes in pH.[4] By definition a chelate involves not only a coordinate covalent bond coming from an amino group, but there is also the possibility of an ionic bond.[5]

In Chapter 8 it was shown that if the dipeptide cell membrane carrier protein, glutathione, is attached to the dipeptide at the dipeptide's terminal carboxyl group instead of the terminal amino group, membrane transport of the dipeptide does not occur.[6] This is possibly due to the inability of γ-glutamyl cyclotransferase to break the carboxyl bond in the membrane transport of dipeptides.[7,8]

By extension, could it not be possible that within the mucosal cell the cleavage of a dipeptide-like amino acid chelate is more difficult because of its use of carboxyl bonds? Is it not possible that peptide chelates can arrive in the plasma from the mucosa intact just as some molecules of protein enter the portal blood as proteoses, peptones and various sizes of polypeptides?

Thus, the purpose of this final study was to ascertain if hydrolysis of the dipeptide-like amino acid chelate within the mucosal cell is essential to its absorption into the serosa, or plasma or if it can be absorbed as an intact chelate. If the latter takes place, then obviously hydrolysis of the chelate occurs at other points in the body where either the amino acid ligands or the metal ion is required.

To elucidate this question, six nonfasted adult malt Sprague-Dawley rats maintained in identical wire bottom cages were chosen for the experiment. Choices were based on similarity of the rats to each other. They had been fed Purina Laboratory Rat Chow *ad libitum* up to the day of experimentation. All of the rats weighed the same, plus or minus 15 grams.

Each rat was anesthetized with ether. An incision was made at midline to expose the intestines. The ileum was located and 15 centimeters of it removed beginning 5 centimeters proximal to the ileocecal valve. The ileum was selected for the experiment because previous studies had shown that maximum peptide absorption occurs in this portion of the small intestine.[9] After removal of the desired portion of the intestine, each rat was sacrificed by cervical dislocation.

The removed ileum was everted and washed three times in Krebs-Ringer Bicarbonate Buffer (pH 7.4) Solution (KRB). This triple washing assured complete removal of luminal contents which may have interfered with the experimental results. Each end of the intestine was tied at both ends with 000 silk thread. During these procedures the intestinal segments were maintained in containers of

KRB Solution into which an oxygen-carbon dioxide mixture (95% O_2 and 5% CO_2 was bubbled. Finally, each intestinal loop was transferred to a vessel similar to that shown in Figure 10-1. The oxygen-carbon dioxide gas mixture continued to be bubbled into these vessels.

Isotonic serosal solution was injected into each intestinal loop by means of an automatic 50cc syringe. The serosal perfusate was collected with a fraction collector. The water bath surrounding the vessel was maintained at 37.5°C to simulate body temperature.

Dipeptide-like amino acid chelates of ^{59}iron were made using either ^{14}C-methionine or ^{14}C-lysine as the amino acid ligands. Pure amino acids were used in the chelates. Each chelate contained only one type of amino acid. The ratio of ^{59}iron to ^{14}C-methionine, based on corrected counts per minute of ^{59}iron and the ^{14}C of the methionine, was 1:1.45 (metal to amino acids). The ratio of ^{59}iron to ^{14}C of the lysine was 1:2.7. In forming the chelates, however, the molar ratios of metal to ligand was exactly 1:2 in both instances. Thus, in each case a dipeptide-like molecule was formed. Ten microliters of the chelate were added to each liter of mucosal solution. In all cases, the pH of the mucosal solution was maintained at 7.1.

Hourly samples of perfusate from the serosal solution were removed and assayed by liquid scintillation. These measurements continued for five hours.

At the conclusion of the experiment each loop was divided into three parts: the proximal, the middle, and the distal sections. The eighteen samples were weighed and then digested with Soluene-350. They too were analyzed for ^{59}iron and ^{14}carbon in the amino acids by liquid scintillation using a program provided by Packard Instruments after appropriate quenching curves were run.

Tables 13-1 and 13-2 summarize the analytical data developed from the chelate incubations. All data are expressed as means of the individual findings.

As Table 13-1 indicates, when the iron was chelated to two amino acids of methionine, there was greater hydrolysis of the chelate in the mucosal cells than when the iron was chelated to two molecules of lysine. The ratio of methionine to iron increased 3.2 times (1.4 to 4.7) in the mucosal tissue and 7.1 times (1.4 to 10.05) in the serosal solution compared to the original ratio in the mucosal solution.

Table 13-1.

MEAN MUCOSAL UPTAKE AND TRANSFER OF ^{59}IRON CHELATED TO ^{14}CARBON IN METHIONINE				
Mean Ratio of ^{59}Iron to ^{14}C-Methionine				
Mucosal Solution	Mucosal Tissue		Serosal Solution	
1:1.4	Proximal	1:4.7	Hour 1	N/A
	Middle	1:4.9	Hour 2	1:7.9
	Distal	1:4.5	Hour 3	1:10.5
			Hour 4	1:11.5
			Hour 5	1:10.2
Mean 1:1.4 S.D. ±		1:4.7 ± .2		1:10.05 ± 1.53

On the other hand, as indicated by Table 13-2, when iron was chelated to lysine, the ratio of metal to amino acid remains relatively constant from the mucosal solution—to the mucosal tissue—to the serosal tissue. There was probably a small amount of hydrolysis of the dipeptide-like lysinate chelate in the mucosal tissue as suggested by the small mean increase of lysine in the tissue (2.7 to 3.05) and the slight decrease of lysine in the serosal solution (2.7 to 1.56), but the majority of the chelate appeared to have been delivered to the serosal solution as an intact chelate or as a free amino acid and free metal with the same ratio, which is less probable if one considers the results of the previous study in Chapter 12.

A close comparison of the two tables also suggests a different rate of speed for mucosal uptake of the iron based on the ligands to which it is chelated. When it is bound to the methionine amino acids, its absorption appears to be more rapid than when it is chelated to lysine.

In a study published by MacInnis, *et al.*,[11] it was determined that amino acids are absorbed from the intestine from at least four specific sites, including sites for neutral amino acids, acid amino acids,

Table 13-2.

MEAN MUCOSAL UPTAKE AND TRANSFER OF ^{59}IRON CHELATED TO ^{14}CARBON IN LYSINE				
Mean Ratio of ^{59}Iron to ^{14}C-Lysine				
Mucosal Solution	**Mucosal Tissue**		**Serosal Solution**	
1:2.7	Proximal	1:2.7	Hour 1	1:1.5
	Middle	N/A	Hour 2	1:2.7
	Distal	1:3.4	Hour 3	1:1.3
			Hour 4	1:1.2
			Hour 5	1:1.1
Mean 1:2.7 S.D. ±		1:3.05 ± .49		1:1.56 ± .65

and basic amino acids. Methionine competed well for all of the sites on the intestinal mucosal cell membrane and was the most rapidly transported of all amino acids studied.

While dipeptides are absorbed at different sites than individual amino acids,[12] Graff[13] has hypothesized that competition for uptake occurs between peptides and dipeptide-like chelates. This, supported by studies which show the intestinal absorption pathway for peptides, is utilized by amino acids.[14] The work of MacInnis, *et al.*,[11] did not distinguish between the absorption sites of amino acids and peptides. The pure peptides, like the amino acids, may have specific sites for absorption. It is possible then, that a pure methionine dipeptide may be absorbed at all of the peptide absorption sites and with greater rapidity than any of the other pure dipeptides. Obviously a mixed dipeptide could make the absorption of dipeptide more complicated than the absorption of amino acids, but mechanism and the affinities for different sites by both types of molecules may be comparable.

If they are the same, then the hypothesis of Graff is probably valid.[13] That would help to explain the apparent differences in the rates of uptake of iron methionate shown in Table 13-1 compared with iron lysinate in Table 13-2. The iron methionate was absorbed at all of the dipeptide absorption sites whereas iron lysinate was absorbed at only the lysine-lysine preferring locus.

Once the iron methionate chelate was absorbed into the mucosal cells, its high degree of dissociation in the tissue was possibly due to incorporation of methionine into protein and a relatively high affinity of methionine for all free amino acid influx and efflux membrane loci. That affinity, by the law of mass action, would have encouraged the hydrolysis of the methionine chelate to amino acids and ionic iron.

There is another consideration to be made regarding the phenomenon of intracellular hydrolysis: Different amino acids have variable stability constants with the same ion depending upon the pH to which they are subjected.[10,15] It has previously been noted that there are measurable changes in cellular pH at the terminal web and as the basement membrane is approached. Since these changes are sufficient to induce oxidation and reduction of ions absorbed through the ionic pathways, they could also possibly lower the stability constant of the methionine chelate to a level that hydrolysis of that chelate could then be facilitated. Rapid movement of free methionine from the mucosal cells to the serosa could then occur.[11]

When the pH is measured, it dropped from 7.1 in the mucosal solution to 6.4 to 6.1 in the tissue. That change in pH would affect the stability of the chelate by encouraging a weakening of the bonds because the more alkaline the pH, the greater the stability of the chelate.[1,2] The lowering of the pH, which results in greater acidity, has the opposite effect.

When these changes in cellular pH are considered with chelates consisting of different ligands, the stability constants may vary enough to facilitate a cleavage of the methionine-iron chelate bonds as compared to the lysine-iron chelate bonds. Thus, the iron lysinate, being more stable, tends to survive the mucosal cell hydrolysis more completely than the iron methionate.

This may not necessarily apply to chelates delivered into tissue cells from the plasma. Movement through mucosal cells is very rapid compared to movement within other body cells. With the relatively methodical transport into tissue cells, there would be a longer period of time in which chelate hydrolysis could occur.

In the previous chapter evidence was presented that indicated that if mucosal cell hydrolysis of the dipeptide-like chelate occurred at all, it was probably in the basement membrane region. The data from the lysine chelate tends to support this observation. As Table 13-2 shows, there is a slight mean increase in amino acid content in the mucosal tissue with a corresponding decrease in the serosal solution.

As postulated above, it appeared that the hydrolysis of the dipeptide-like amino acid chelate tended to increase in relation to the time in the cell. Assuming stability constants of the chelates are comparable, the ratio of ligand to metal in the serosal solution of the lysine chelate indicated that this may be what happened. These ratios also suggested that perhaps cleavage of only one molecule of amino acid from the chelate could have occurred. The mean ratio of the ligand to the metal suggested that some of the serosal chelates may have been single amino acid chelates rather than dipeptide-like amino acid chelates. The ratio of ligand to metal in the second hour indicated that intact dipeptide-like amino acid chelates could be, and were, passed from the mucosa to the serosa without hydrolysis.

From this, as well as the studies presented in Chapters 9 through 12, it is postulated that the dipeptide-like amino acid chelates are absorbed into the mucosal cell intact via the dipeptide pathway and not the ionic pathway. Whereas the latter uses a pump similar to the sodium-potassium pump and a two-step process for active uptake, the dipeptide-like amino acid chelate may depend upon a different series of enzymatic reactions of absorption. When the mineral is chelated with two amino acids, the resulting molecule's absorption occurs much like that of a dipeptide. Thus, the intestines treat it as such by avoiding brush border hydrolysis, absorbing it intact, with the metal transversing the mucosal cell membrane as a moiety of the original chelate molecule.

REFERENCES

1. Bell, C., *Metal Chelation Principles and Applications* (Oxford: Clarendon Press), 11, 1977.
2. Mellor, D., "Historical Background and Fundamental Concepts," in Dyer, F. and Mellor, D., eds., *Chelating Agents and Metal Chelates* (New York: Academic Press), 45, 1964.

3. Guyton, A., *Textbook of Medical Physiology* (Philadelphia: W.B. Saunders Co.), 768, 1971.
4. White, A., *et al.*, *Principles of Biochemistry* (New York: McGraw Hill), 108-112, 1973.
5. Mellor, op. cit., 1.
6. Matthews, D. and Burston, D., "Intestinal Transport of Peptides," in Kramer, M. and Lauterbach, F., eds., *Intestinal Permeation* (Amsterdam: Excerpta Medica), 136, 1977.
7. Meister, A., "On the Enzymology of Amino Acid Transport," Science, 180:33, 1973.
8. Tate, S. and Meister, A., "Interaction of Gamma-Glutamyl Transpeptidase with Amino Acids, Dipeptides, and Derivatives and Analogues of Glutathione," J. Biol. Chem., 249:7593, 1974.
9. Rhodes, J., *et al.*, "Intestinal Hydrolysis of Dissacharides and Peptides: Comparison of Hydrolysis and Perfusion Studies," in Elliott, K. and O'Connor, M., eds., *Peptide Transport and Hydrolysis* (Amsterdam: Elsevier), 245-259, 1977.
10. Bjerrum, J., *Metal Ammine Formation in Aqueous Solution* (Copenhagen: Haase), 1941.
11. MacInnis, A., *et al.*, "Specificity of Amino Acid Transport in the Tapeworm Hymenolepis Diminuta and its Rat Host," Rice University Studies, 62:183, 1976.
12. Rubino, A. and Guandaline, S., "Dipeptide Transport in the Intestinal Mucosa of Developing Rabbits," in Elliott, K. and O'Connor, M., eds., *Peptide Transport and Hydrolysis* (Amsterdam: Elsevier), 61, 1977.
13. Graff, D., "Radioactive Isotope Research with Chelated Minerals," in Ashmead, D., ed., *Chelated Mineral Nutrition in Plants, Animals and Man* (Springfield: Charles C Thomas), 277, 1981.
14. Munro, H. and Crim, M., "The Proteins and Amino Acids," in Goodhart, R. and Shils, M., eds., *Modern Nutrition in Health and Disease* (Philadelphia: Lea and Febiger), 67, 1980.
15. Jensen, N., "Biological Assimilation of Metals," U.S. Patent No. 4,167,564, Washington, D.C., Sept. 11, 1979.

CHAPTER 14

SUMMARY AND IMPLICATIONS

WHEN a mineral is ingested, there is more than one transport system that carries it across the intestine depending upon the form of the mineral entering the intestine. If it is presented to the mucosal membrane as a soluble salt, it will be absorbed as a cation by active transport and to a lesser extent, diffusion. If it is presented as a dipeptide-like amino acid chelate, its uptake from the mucosa follows another unrelated pathway[1,2] which also depends on active transport and some diffusion.

A mineral salt, after ingestion is solubilized and ionized in the acid pH of the stomach.[3] There are many chemical reactions in the gut that can precipitate the cation as it enters the small intestine. If none of these occur, the mineral is presented to the integral proteins in the brush border of the duodenum.[4] While uptake of the cation can probably occur any place in the small intestine, it is in the relatively low pH of the duodenum that the ion tends to maintain solubility.[3] With the addition of the pancreatic buffer, enzymes, and bile, the pH of the small intestinal milieu rises. It is then that precipitation of the minerals still unabsorbed and in the luminal cavity of the gut can occur.[3]

Optimally, for intestinal absorption of the cation to occur, it must generally take place shortly after the solubilized minerals have passed the pylorus and entered the duodenum.[4] It is there while still in the ionic state the minerals are chelated with the integral proteins which transverse the membrane of the mucosal cell.[5]

The mucosal cell membrane consists of two opposing monomolecular layers of phosphoglyceride molecules with their non-moisture absorbing hydrophobic regions aligned towards each other (Figure 3-10). Their water-absorbing hydrophilic groups are oriented towards the exterior and interior of the mucosal membrane, but also extrude from it. The protruding parts have ionic residues (+ and −) of the protein on their surfaces. It is these ionic residues that allow the integral protein to chelate the cation presented to it.[6,10]

The chelate formed is not dipeptide-like in physical structure. Instead, the metal ion is attached by coordinate covalent bonding to the terminal end of the integral protein.[7,8,9] The mineral is absorbed against a concentration gradient. Therefore, the process is endergonic and consumes adenosine triphosphate (ATP) in the plasma membrane which was probably manufactured in the mitochondria of the mucosal cell.[11] The phosphatase enzyme thought to be associated with supplying energy for the transport process is concentrated in the microvilli.[6]

As the mineral enters the mucosal cell it is subjected to a pH change.[12,13] This causes a change in the stability constant of the integral protein chelate.[14] There is competition between the integral protein and peptide groups called carrier proteins within the cell for the metal ion. As the stability constant is lowered in the integral protein chelate, the attraction for the carrier proteins is sufficient to induce the cation to leave the integral protein in favor of the carrier protein. As this occurs there may also be a change in the valence of the metal such as the oxidation of iron from the ferrous (+2) state to the ferric (+3) state.[13]

Through the process of chelation the carrier protein collects several atoms of a metal and then, by a means which is still unknown, transports them to the basement membrane of the mucosal cell.[15,20]

At the basement membrane there is still another change in cellular pH.[13] This probably induces the carrier protein to give up its cargo of ions due to changes in stability constants of the carrier protein chelate. Again, there is probably a valence change in the metal ions.[13] In the case of iron, it is reduced to the ferrous state.

This change in valence is probably essential to induce integral proteins embedded in the basement membrane to chelate the cations and carry them across that membrane to the serosa or plasma. The mechanism by which the movement across the basement membrane is accomplished is probably the same as that in the brush border membrane. A change in the pH at the exterior side of the basement membrane would induce the integral protein to release the cation due to changes in the chelate's stability constant.[14]

For the above mineral absorption mechanism to operate within the intestine, the mineral must remain in the ionic state.[3] If the cation loses its solubility property, it cannot be chelated by the integral proteins in the mucosal cell membrane.[6,10] Unless the ion is attached to this protein, it cannot be absorbed by the active transport mechanism just described.[11]

Nevertheless, there is at least one other active transport mechanism located in the intestine that is not dependent on ionization of the metal in the gut. Indeed it functions only if the metal ion is chelated with two amino acids to form a dipeptide-like molecule prior to its presentation to the intestine.[1,2] In this form the mineral is absorbed not as an ion but as part of a peptide molecule because it occupies a central position between the two amino acids which together with the metal form the chelate. The amino acids "smuggle" the metal into the mucosal cell.[21,22]

The formation of the dipeptide-like amino chelate occurs when the divalent cation is ionized. The amino acid ligands contain a negative charge through the removal of hydrogen. When the ligands and the ion are introduced to each other in the presence of pH changes, chelation can occur.

Chelation involves formation of coordinate covalent bonds between each donor ligand and metal. The covalent bond comes from the donation of electrons to the metal from the amino, carboxyl, and/or sulfurhydral groups of the ligand. In conjunction with the covalent bonds there can also be ionic bonds. In a dipeptide-like amino acid chelate there are ten bonds forming two heterocyclic rings. The metal occupies the central position between the two amino acids which forms the chelate. This is seen in Figure 14-1.

A heterocyclic ring is formed between the metal and each amino acid in the chelate.[9,23] The ring is formed by at least five bonds, two of them involving the metal. A chelate with less than five bonds does not result in a stable chelate ring.[24]

A double heterocyclic ring is formed when two amino acids form a chelate with the same metal.[9] The attachment of the second amino acid ligand results in a dipeptide-like amino acid chelate as opposed

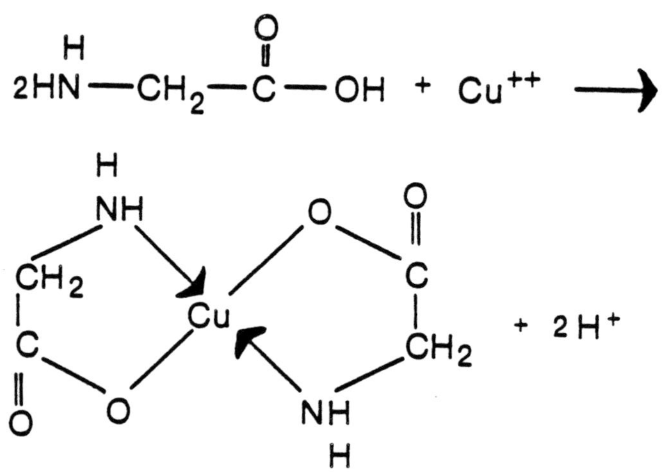

Figure 14-1. The formation of a dipeptide-like chelate of copper. Coordinate covalent bonds come from the amino group and ionic bonds from the carboxyl group.

to an amino acid chelate which involves the chelation of the metal with just one amino acid. Our studies have suggested that the amino acid chelate may not be as stable *in vivo* as the dipeptide-like amino acid chelate due to greater exposure of the metal in a single ring. Both types of chelates have essential roles in the body as seen in the chelation of the ion by the integral protein which forms a molecule similar to the amino acid chelate.

Chelates can be formed with several amino acids or with polypeptides. Nevertheless, only two amino acids of a dipeptide-like chelate with a total molecular weight including the metal of under 1500 is able to penetrate the cell membrane without requiring additional luminal hydrolysis.[25] That means that while two peptide chains can chelate an ion, the resulting molecule is too large to be absorbed intact. As seen earlier, only a small portion of the tripeptides survive luminal hydrolysis, whereas dipeptides do. Thus, a molecule that is larger than a dipeptide cannot effectively smuggle the ion into the mucosal cell. If membrane digestion does occur, as in the case of larger chelate molecules,[26] the bonds between the metal and the ligand may inadvertently be broken too. If this occurs, the potential value of the chelate as a secondary transport medium for the metal is lost. Even if the chelate molecule is small—that is, under a molecular weight of 1500—generally it will still not survive the acid pH of the stomach as an intact molecule unless the chelate is stabilized in the manufacturing process.[27]

If this type of dipeptide-like amino acid chelate is formed, it will survive the peptidase enzymes of the gut that break the interior peptide bonds[27,28] possibly due to the presence of the cation in the molecule. It will also survive the acid pH of the stomach.

Ultimately, the surviving dipeptide-like amino acid chelate passes through the pylorus to the small intestine. The rate of transfer from the stomach to the intestine is dependent upon the quality, the quantity, and digested state of the food in the stomach.

MacInnis, *et al.*,[29] suggest that most, if not all, mucosal cells of the small intestine are capable of absorbing amino acids although the majority of this absorption occurs below the pancreatic duct and is completed in the upper one-half of the small intestine.[30] In all probability dipeptides and dipeptide-like amino acid chelates have similar absorption patterns. Based on observations presented in Chapter 11, it appears the latter attain maximum absorption in the small intestine beyond the ionic absorption sites of the duodenum.

When the dipeptide-like amino acid chelate is presented at the brush border membrane of the mucosal cell, it is believed to be coordinately attached by its terminal amino group to glutathione, a tripeptide found in the cell membrane.[31,32] The enzymatic breakdown of the glutathione results in the movement of the dipeptide-like amino acid chelate from the luminal side of the mucosal membrane to the cytoplasm within the cell.

As shown in Figure 14-2, the glutathione is thought to be attached to the dipeptide-like amino acid chelate at the terminal amino group (NH_2) of one of the two amino acids forming the chelate. If a carboxyl group is substituted for the amino group, the transport of the chelate is abolished.[33] This terminal amino group of the chelate is believed to be attached to the gamma-glutamyl linkage of the tripeptide in the mucosal cell membrane.[34]

The resulting chelate-glutathione compound would then be degraded to gamma-glutamyl dipeptide-like amino acid chelate and cysteinyl-glycine by the enzyme gamma-glutamyl cyclotransferase. The next step would be the cleavage of the gamma-glutamyl dipeptide-like amino acid chelate by the enzyme gamma-glutamyl cyclotransferase. This would degrade the gamma-glutamyl chelate into 5-oxy-L-proline and the dipeptide-like amino acid chelate.[31,32,37,38,39]

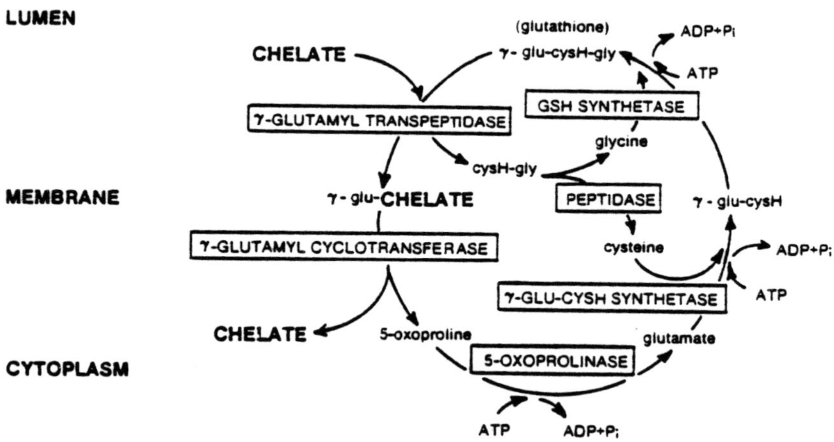

Figure 14-2. The proposed membrane transport of a dipeptide-like chelate.

At this point the dipeptide-like amino acid chelate would be in the cytoplasm of the mucosal cell. The active transport of the chelate from the lumen to the cytoplasm would have been accomplished in two enzymatic steps. The involvement of ATP would be necessary to restore glutathione to carry the next molecule of the dipeptide-like amino acid chelate (the substrate) across the brush border into the cell.

The final release of the dipeptide-like amino acid chelate into the cytoplasm would occur at the terminal web as a result of enzymatic action at the original amino group linkage point. Some of the chelate would then move across the cell to the basement membrane and into the plasma as intact molecules. At present no intracellular carrier has been identified which functions to move chelates through the mucosal cell. Perhaps osmotic influx pressure and the kinetic energy of diffusion may account for the movement of the dipeptide-like amino acid chelate to the basement membrane from the luminal surface.

A certain amount of peptide hydrolysis can occur within the cell.[38] As far as intracellular digestion of dipeptide-like amino acid chelates is concerned, this probably occurs less frequently than intracellular dipeptide hydrolysis because of the blocking action of the cation in the chelate molecule. The actual cleavage of the ion from amino acid ligands appears to take place in the basement membrane where there is a shift in pH.[39] If hydrolysis occurs, the coordinate bonds are probably broken by enzymatic action in combination with acids which have lowered the stability constant and favored the release of the cation. Any ionic bonds would be broken by the change in pH.

If hydrolysis of the dipeptide-like amino acid chelate occurs, it appears that frequently only one ligand is removed. The metal remains attached to the other amino acid. An amino acid chelate results and is absorbed into the plasma as such.

If the metal is released as a free ion, it is probably chelated by the peripheral proteins in the basement membrane and transported across the membrane to the serosa, or plasma. This would occur by a mechanism that is similar to that for any equivalent ion being absorbed in the ionic state, the details of which were described earlier. Reported isotope studies indicated that the freeing of the metal to the ionic site does not generally occur when it is absorbed as a dipeptide-like amino acid chelate, such as a metal lysinate.

Intracellular hydrolysis of the dipeptide-like amino acid chelate is not an essential requirement for its passage out of the mucosal cell into the serosa or plasma. Efflux can occur as an intact chelate just as it can for proteoses, peptones, and small polypeptides.[40] Hydrolysis seems to be somewhat dependent upon the amino acid ligands used to form the chelate, perhaps the molar ratio of metal to ligand in the mucosa, and the length of time the peptide chelate remains in the mucosal cell.

In either case, where the chelate either does not undergo complete intracellular hydrolysis or remains totally intact (so that the metal is bound to one or more of the original amino acids), efflux probably depends on the binding of a terminal amino group of one of the ligands to glutathione in the basement membrane and moving it through the basement membrane.[34] Thus, the exit of the chelate from the mucosal cell to the serosa or plasma would be accomplished by active transport.

This *secondary process* of mineral absorption has three distinct advantages over ionic mineral absorption. First, it is presumably more rapid because the two-step absorption model required of ions is avoided. Second, its uptake and absorption are greater because a less critical intestinal environment from the standpoint of gut pH, precipitating ligands, etc., is required for absorption. Third, and this evolves out of the second advantage, more predictive absorption of a mineral within the bounds of peptide absorption may be possible if its chelation process is standardized.

The implications of these advantages are far reaching and are as flexible as mineral nutrition. *In vivo* isotope studies have shown that not only do dipeptide-like amino acid chelates of iron and zinc result in higher blood levels of those minerals as a function of time,[44,45] but also those of calcium[41,45] and manganese.[42,45] Copper has not been tested in the isotopic form and the half life of an isotope of magnesium precludes it from this type of study, but *in vitro* research on copper[1] and *in vivo* studies of magnesium[43] suggest a mechanism for them similar to those which exist for iron, zinc, calcium, and manganese.

To illustrate the efficacy of supplementation with chelates compared with salts in mineral deficient animals, twenty two-day old male pigs were given ^{59}Fe in the form of ferrous chloride or iron dipeptide-like amino acid chelate. Hemoglobin samples were obtained every hour beginning one hour after the initial oral dose and continuing for twenty hours. During the period from birth to the end of this experiment the pigs received no additional nutrition. Consequently there was no gastro-intestinal interference with the intestinal absorption of the iron. Figure 14-3 shows the mean level of ^{59}Fe in the hemoglobin of each group of pigs.[45,46]

This study helped determine that the majority of the dipeptide-like iron chelate was absorbed into the plasma as intact molecules. Consequently, most of it was taken directly to the spleen rather than first going to the liver as would have been expected in the case of the inorganic iron salt.[46]

A major role of the spleen is to disintegrate old red blood cells and remove the iron for production of new hemoglobin. The molecular structure of the dipeptide-like amino acid chelate is close enough to the iron removed from the red blood cells to allow the ingested and absorbed chelate to be taken directly to the spleen rather than first being stored in the liver and restructured for transfer to the spleen later. From the spleen the iron moves to the bone marrow for production of new red blood cells. This can be interpreted from Fig-

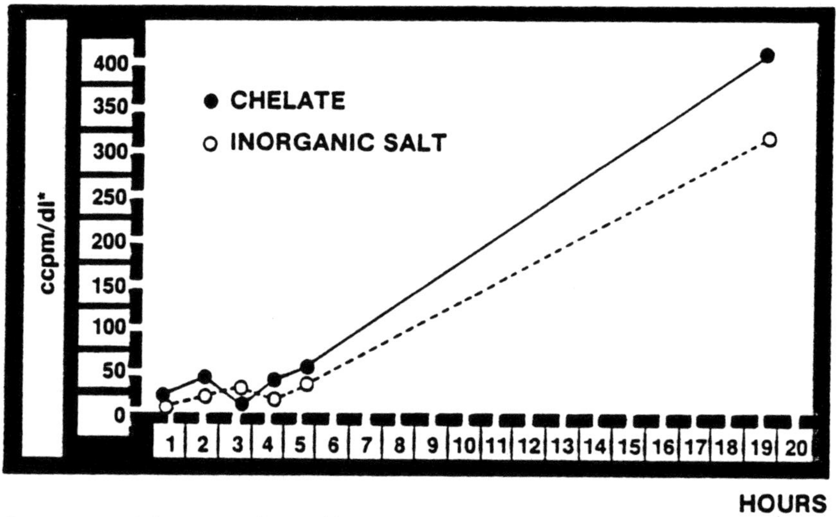

*ccpm = corrected count per minute, with count corrected for background radiation.

Figure 14-3. Hemoglobin level of ^{59}Fe after oral administrtion of the chelated and inorganic salt forms.[8]

ure 14-3[8] particularly in light of the following: "In a study of incorporation of Fe in RBC, six littermate male pigs about twelve hours old were given either amino acid chelated Fe or $FeCl_2$. They had not nursed or received any other postnatal nutrition other than the two forms of ^{59}iron. In five hours the Fe content of the RBC of the piglets receiving the chelate was 32% higher than for the control group which had received equal amounts of Fe as $FeCl_2$. The Fe content of the RBC of the control group dropped over 25% in five hours suggesting that even though this form of Fe may have been absorbed, it was not being metabolized."[46] This is seen in Figure 14-4.[46]

The potential use of chelates becomes even more meaningful if one considers the study of Yamamoto, *et al.*,[47] In this study twelve gestating sows all receiving a normal diet were randomly divided into three groups. Eight of the pigs were fed 600 mg of iron daily in their gestation rations as either ferrous fumarate or dipeptide-like iron amino acid chelate. Four sows receiving a regular gestation diet but no supplemental iron were used as a control group. The serum ferritin concentration in the sows was measured four weeks prior to farrowing, at farrowing, and at weaning. The reason ferritin concentration was measured is because it has been established to be the indicator of hematopoetic status in the body.[48] Table 14-1 presents the summarized findings.[47]

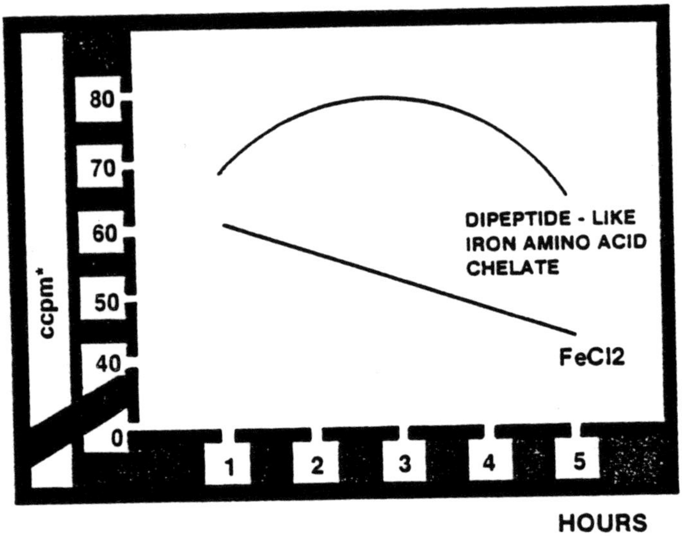

*ccpm = corrected count per minute, with count corrected for background radiation.

Figure 14-4. Level of ^{59}Fe in the RBC after oral administration of the chelated and inorganic salt forms.[46]

Table 14-1[47]

SERUM FERRITIN CONCENTRATION IN SOWS (ng/ml)				
	NUMBER OF SOWS	4 WEEKS PRIOR FARROWING	AT FARROWING	AT WEANING
CONTROL	4	7.3 ± 7.1	2.0 ± 2.6	3.0 ± 2.6
DIPEPTIDE - LIKE IRON AMINO ACID CHELATE	4	13.0 ± 9.6	16.5 ± 10.4	11.5 ± 8.3
FERROUS FUMERATE	4	7.3 ± 6.7	4.5 ± 3.7	5.3 ± 4.7
(mean + SD)				

This study also showed statistically significant differences in the hemoglobin concentration, the hematocrit values, and the serum ferritin concentration of their piglet offspring which are commonly deficient in those parameters at birth. Tables 14-2, 14-3 and 14-4, and Figures 14-5, 14-6 and 14-7 present these data.[47]

Table 14-2[47]

HEMOGLOBIN CONCENTRATION IN PIGLETS (g/dl)					
GROUP \ WEEK	0	1	2	3	4
CONTROL	8.5	7.3	6.2	6.0	6.0
DIPEPTIDE - LIKE IRON AMINO ACID CHELATE	9.6	8.0	7.6	7.8	8.5
FERROUS FUMERATE	9.7	7.5	6.6	6.7	7.4

Table 14-3[47]

HEMATOCRIT VALUE IN PIGLETS (%)					
GROUP \ WEEK	0	1	2	3	4
CONTROL	30.5	25.4	22.7	22.2	23.3
DIPEPTIDE - LIKE IRON AMINO ACID CHELATE	32.4	28.0	27.5	28.2	30.6
FERROUS FUMERATE	33.2	26.3	24.0	24.9	27.8

The explanation for the improved blood picture of the offspring whose gestating mothers received the dipeptide-like iron amino acid chelate is found in a paper presented by Ashmead and Graff at the

Table 14-4[47]

SERUM FERRITIN CONCENTRATION IN PIGLETS (ng/ml)						
	Number of Piglets	0 Week	1 Week	2 Week	3 Week	4 Week
CONTROL	12	36.3 ± 25.3 [a]	34.8 ± 32.5 [d]	9.5 ± 9.1	10.3 ± 9.8	8.8 ± 8.2
DIPEPTIDE - LIKE IRON AMINO ACID CHELATE	12	157.2 ± 88.0 [b]	98.2 ± 75.0 [e]	7.2 ± 2.7	17.4 ± 25.9	13.9 ± 13.7
FERROUS FUMERATE	11	20.1 ± 12.1 [c]	36.1 ± 41.8 [f]	6.8 ± 7.1	6.7 ± 8.5	7.5 ± 7.5

a–b = $p < .001$
b–c = $p < .01$
d–e, e–f = $p < .05$

(mean ± SD)

1982 International Pig Veterinary Society Congress.[49] Twelve Sprague-Dawley female rats that had been synchronously impregnated, were randomly divided into four groups. Four days prior to calculated parturition each female received 4.4 microcuries of ^{59}Fe either as the dipeptide-like iron amino acid chelate or as $FeCl_3$ with or without a cholinesterase inhibitor. The dose was combined with 25 μl of water and administered by automatic pipette. Necropsy and removal of the fetuses occurred one day before calculated parturition. Whole body assays for ^{59}Fe in each fetus was conducted. The results are seen in Table 14-5.[49]

This study suggests that there is more than one mechanism involved in the transport of maternal iron to the fetus. If there was total digestion of the dipeptide-like iron amino acid chelate in the mucosal cells or earlier in the gut prior to its uptake into the plasma, then all iron delivered to the fetuses would have been maternally carried by transferrin. It would have had a similar molecular weight instead of the disproportionate 500,000 for ferritin and less than 1500 for the chelate. If all of the iron were bound to maternal transferrin in maternal plasma, regardless of source, it would have been transferred from the maternal blood through the chorionic villi of the areolae bound to specific protein molecules which had been produced by the placenta to specifically transport the iron from the mother to the fetus.[50,51]

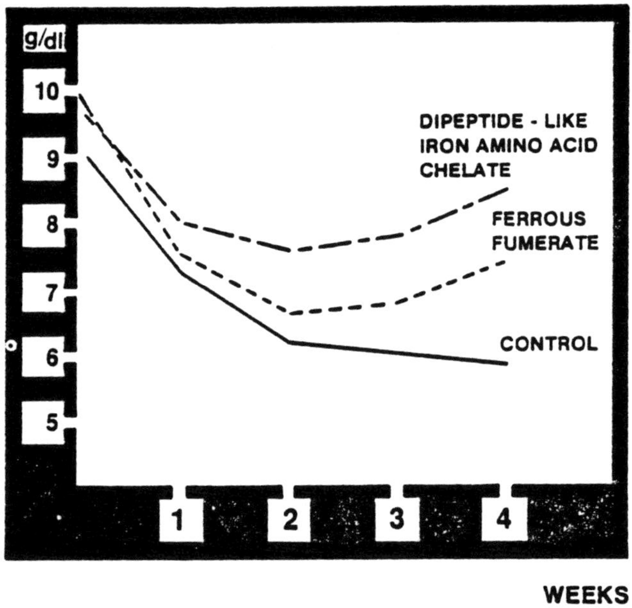

Figure 14-5. Changes in hemoglobin concentration of piglets.[47]

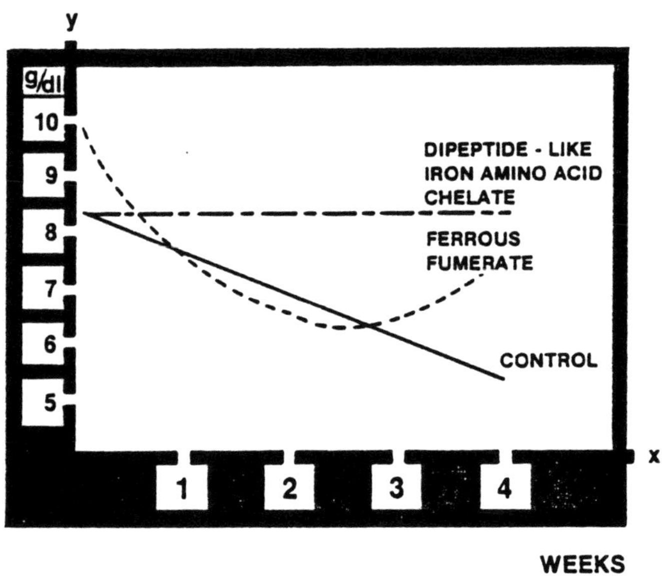

Figure 14-6. Regressive equations of hemoglobin concentration of piglets.[47]

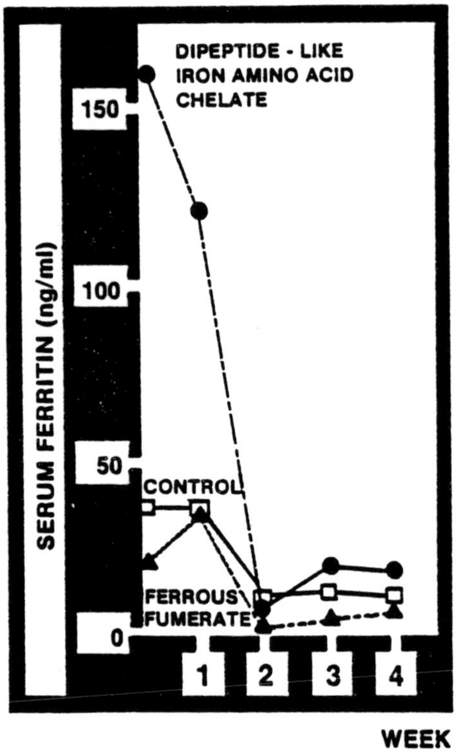

Figure 14-7. Changes in serum ferritin concentration of piglets.[47]

Now referring to Table 14-5, "If fetal uptake of iron from maternal plasma was due solely to placentally-produced carrier proteins, then the iron from both the chelate and the chloride groups would have had equal affinity for the carrier for iron uptake regardless of the form of iron administered. In this case, increased uptake from the chelate group would have been due solely to increased maternal intestinal absorption of that form of iron; but when the results of the chelate and chloride groups are compared and the vasodilator added, the increased fetal uptake of iron is not proportional. If these comparisons were proportional, a similar placental pathway for both forms of iron, dependent upon the same number of carrier proteins, would be indicated."[49]

This dissimilarity suggests that the dipeptide-like iron amino acid chelate enters the maternal plasma as an intact molecule. It has a molecular weight of less than 1500, and is therefore small enough

Table 14-5.

Treatment	cc/m/Fetus	
	Mean	**S.D.**
$^{59}FeCl_3$	90.3	± 28.6
^{59}Fe A.A. Chelate	127.0	± 79.6
$^{59}FeCl_3$ + 2,2-dichlorovinyl dimethyl phosphate	111.1	± 26.2
^{59}Fe A.A. Chelate + 2,2-dichlorovinyl dimethyl phosphate	293.5	± 157.7

to be taken into fetal circulation as an intact molecule without undergoing the changes outlined above.

This study not only lends support for multiple absorption mechanisms for the intestinal uptake of minerals into the plasma, but also supplies a far-reaching benefit that can accrue from this second pathway utilized by dipeptide-like amino acid chelates in treatment of a mineral deficiency.

In studies reported by Rajan, *et al.*,[52] copper and zinc chelates of L-Dopa (phenylalanine analogs) were injected intraperitoneally. One hour later the test animals were sacrificed, their brains removed and assayed for L-Dopa. These researchers found increases in brain levels of L-Dopa up to 150% of controls which received unchelated L-Dopa. This is significant because the brain uses Dopa in the synthesis of epinephrine and norepinephrine.[53] Furthermore, L-Dopa is actively transported to the brain through the blood brain barrier. Because it can cross the blood brain barrier, this molecule is frequently used for the transport of nutrients into the brain tissue.[54]

Similar studies[54] using dipeptide-like ^{65}zinc amino acid chelates or ^{65}zinc chloride instead of the phenylalanine analog ligand showed percentage increases from an intraperitoneal dose almost identical to what was reported by Rajan, *et al.*, The zinc chelated with two amino acids of phenylalanine is transported across the blood brain barrier in amounts 150% that of controls receiving unchelated zinc chloride.[54] Thus, this dipeptide-like chelate can be used to carry nutrients to the brain.

The benefits resulting from the chelation of calcium ions to amino acids in a molar ratio of 1:2, respectively, is equally beneficial. In test animals in which 14 microcuries of ^{45}Ca was orally administered as $CaCl_2$ or as the dipeptide-like amino acid chelate, it was found that one week after the dose the animals receiving the chelate had retained 64% more calcium than the group receiving $CaCl_2$. These data are summarized in Table 14-6.[55]

Table 14.6[55]

MEAN METABOLISM OF CALCIUM IN ANIMALS		
	CONTROL*	EXPERIMENTAL
FRONTAL BONE	3,682	5,772
MASSATER	602	904
GASTROC. (Muscle)	614	1,206
HEART	642	932
LIVER	664	742
R. CEREBRUM	698	804
KIDNEY	686	730
LUNG	676	648
BLOOD SERUM	8.4	31.0
NO. RED BLOOD CELLS	18.6	13.2
WHOLE BLOOD	27.0	44.2

*DATA ARE EXPRESSED AS CORRECTED COUNTS PER MINUTE

The implications for bone development, for improved nerve and muscle functioning, for coagulation of the blood, and for improved eggshell quality in poultry are evident.

Other studies demonstrate the implications for improved mineral nutrition when the essential mineral is ingested as a dipeptide-like amino acid chelate. Space does not allow presentation, and in reality should be examined on a benefit per indication basis rather than as a summary.

Suffice it to say that when the body is provided with essential minerals that have been chelated with two amino acids per metal ion and then stabilized, there appears to be significant improvements in the performance of the system within the body which is affected by the needed minerals compared to equivalent systems receiving minerals from inorganic mineral salt sources. The organism is more capable of reaching its genetic potential due to improved mineral nutrition.

The ability to achieve these benefits is due to the physiological capability of the intestine to absorb minerals either as ions or as the moiety of a molecule that has the appearance of a dipeptide. In the latter form there is significantly greater absorption due to avoidance of interfering substances in the lumen. Furthermore, as a function of time there is also greater absorption possibly due to the avoidance of the saturable two-step absorption mechanisms that operate in the case of ionic absorption. Thus, as a dipeptide-like amino acid chelate, the intestinal absorption and transfer of metal to the plasma is significantly more efficient.

REFERENCES

1. Graff, D., *et al.*, "Absorption of Minerals Compared with Chelates Made From Various Protein Sources into Rat Jejunal Slices *in vitro*," Address to Utah Acad. Arts, Letters, Sci., Salt Lake City, April 1970.
2. Ashmead, H., "Tissue Transportation of Organic Trace Minerals," J. Appl. Nutri., 22:42, 1970.
3. Waldron-Edward, D., "Effects of pH and Counter-Ion on Absorption of Metal Ions," in Skoryna, S. and Waldron-Edward, D., eds., *Intestinal Absorption of Metals, Trace Elements and Radionuclides* (Oxford: Pergamon Press), 381, 1971.
4. Hendrix, T., "The Absorptive Function of the Alimentary Canal," in Mountcastle, V., ed., *Medical Physiology* (St. Louis: C.V. Mosby Co.),, 1147, 1974.
5. Giese, A., *Cell Physiology* (Philadelphia: Saunders College), 192, 1973.
6. White, A., *et al.*, *Principles of Biochemistry* (New York: McGraw Hill), 291-297, 1973.
7. Pauling, L., *The Nature of the Chemical Bond* (Ithaca: Cornell University Press), 61, 1960.
8. Ibid., 6.
9. Mellor, D., "Historical Background and Fundamental Concepts," in Dwyer, F. and Mellor, D., eds., *Chelating Agents and Metal Chelates* (New York: Academic Press), 1-17, 1964.
10. Garay, R. and Farrahan, P., "The Interaction of Sodium and Potassium with the Sodium Pump in Red Cells," J. Physiol., 231:297, 1973.
11. Freeman, J., *et al.*, "Fine Structure of the Intestinal Epithelial Cell in Iron Absorption," in Skoryna, S. and Waldron-Edward, D., eds., *Intestinal Absorp-*

tion of Metals, Trace Elements and Radionuclides (Oxford: Pergamon Press), 395, 1971.
12. Henderson, P., "Ion Transport by Energy-Conserving Biological Membranes," Ann. Rev. Microbiol., 25:393, 1971.
13. Granick, S., "Structure and Physiological Functions of Ferritin," Physiol. Rev. 31:497, 1951.
14. Moeller, T. and Horwitz, E., "Chelation," in Comar, C. and Bronner, F., eds., *Mineral Metabolism* (New York: Academic Press), V1, 109, 1960.
15. Wasserman, R., *et al.*, "Vitamin D-Dependent Calcium Binding Protein: Purification and Some Properties," J. Biol. Chem., 243:3978, 1968.
16. Van Campen, D. and Kowalski, T., "Studies on Zinc Absorption: ^{65}Zn Binding by Homegenates of Rat Intestinal Mucosa," Proc. Soc. Exp. Med., 136:294, 1971.
17. MacIntyre, I., "Discussion on Magnesium Metabolism in Man and Animals," Proc. Royal Soc. Med., 53:1037, 1960.
18. Tansy, M., "Intestinal Absorption of Magnesium," in Skoryna, S. and Waldron-Edward, D., eds., *Intestinal Absorption of Metals, Trace Elements and Radionuclides* (Oxford: Pergamon Press), 199, 1971.
19. Thompson, A., *et al.*, "Interrelationship of Intestinal Transport System for Manganese and Iron," J. Lab. Clin. Med., 78:642, 1971.
20. Underwood, E., *Trace Elements in Human and Animal Nutrition* (New York: Academic Press), 62, 1977.
21. Ames, B., *et al.*, "Illicit Transport, the Oligopeptide Permease," Proc. Natl. Acad. Sci., 70:456, 1973.
22. Mellander, O., "The Nutritional Significance of Some Peptides," Voeding, 16:336, 1955.
23. Bell, C., *Metal Chelation Principles and Applications* (Oxford: Clarendon Press), 9, 1977.
24. Ibid., 16-18.
25. Tiffin, L., "Translocation of Micronutrients in Plants," in Dinauer, R., ed., *Micronutrients in Agriculture* (Madison: Soil Science Society), 207, 1972.
26. Cardon, B., "The Importance of Chelation in Trace Minerals," Paper given to Montana Nutrition Conference, February 11, 1963.
27. Jensen, N., "Biological Assimilation of Metals," U.S. Patent No. 4,167,564, Washington D.C., September 11, 1979.
28. Hendrix, op. cit., V2, 1162.
29. MacInnis, A., *et al.*, "Specificity of Amino Acid Transport in the Tapeworm Hymenolepis Diminuta and its Rat Host," Rice University Studies, 62:183, 1976.
30. Code, C., ed., *Handbook of Physiology* (Baltimore: Williams and Wilkins Co.), V3, 1125, 1968.
31. Meister, A., "On the Enzymology of Amino Acid Transport," Science, 180:33, 1973.
32. Tate, S., and Meister, A., "Interaction of Gamma-Glutamyl Transpeptidase with Amino Acids, Dipeptides, and Derivatives and Analogues of Glutathione," J. Biol., Chem., 249:7593, 1974.

33. Matthews, D. and Burston, D., "Intestinal Transport of Peptides," in Kramer, M. and Lauterbach, F., eds., *Intestinal Permeation* (Amsterdam: Excerpta Medica), 136, 1977.
34. White, op. cit., 633.
35. Meister, A., "Biochemistry of Glutathione," in Greenberg, D., ed., *Metabolism of Sulfur Compounds, Metabolic Pathways* (New York: Academic Press), 101-188, 1975.
36. Meister, A. and Tate, S., "Glutathione and Related γ-Glutamyl Compounds; Biosynthesis and Utilization," Ann. Rev. Biochem., 45:559, 1976.
37. Meister, A., *et al.*, "On the Function of the γ-Glutamyl Cycle in the Transport of Amino Acids and Peptides," in Elliott, K. and O'Connor, M., eds., *Peptide Transport and Hydrolysis* (Amsterdam: Elsevier), 123-143, 1977.
38. Crim, M. and Munro, H., "Protein," in Hegsted, D., *et al.*, ed., *Present Knowledge in Nutrition* (New York: The Nutrition Foundation), 44, 1976.
39. Schepartz, B., *Regulation of Amino Acid Metabolism in Mammals* (Philadelphia: W.B. Saunders Co.), 34, 1973.
40. Greyton, A., *Textbook of Medical Physiology* (Philadelphia: W.B. Saunders Co.), 768, 1971.
41. Ashmead, H., *et al.*, "The Influence of Stability Constants on Calcium Metabolism *in vivo*" Unpublished, 1974.
42. Ashmead, H., *et al.*, "Stability Constants of Manganese[54] Chelates and Their Effects in Manganese Metabolism," Unpublished, 1974.
43. Ashmead, H., and Fouad, T., "Trace Mineral Interrelationships, New Techniques of Detecting Lead and Other Heavy Metals in Animals, and the Role Organic Chelated Trace Elements Play as Enzyme Catalysts," Paper given to Oklahoma Vet. Med., Conv., Tulsa, February 8, 1971.
44. Graff, D., *et al.*, "Radioactive Isotope Research with Dipeptide-Like Amino Acid Chelates," Unpublished, 1980.
45. Beck, B., "Radioactive Isotope Research with Chelated Minerals," in Ashmead, D., ed., *Chelated Mineral Nutrition in Plants, Animals and Man* (Springfield: Charles C Thomas), 152, 1982.
46. Ashmead, D., "Oral Supplementation of Fe and Zn in Pigs," Mod. Vet. Practice, 64:69, 1983.
47. Yamamoto, M., *et al.*, "A Comparative Study of Iron Proteinate and Ferrous Fumerate on Pig Performance," in Ashmead, D., ed., *Chelated Mineral Nutrition in Plants, Animals and Man* (Springfield: Charles C Thomas), 210, 1982.
48. Miles, L., *et al.*, "Measurement of Serum Ferritin by a 2-Site Immunoradiometric Assay," Anal. Biochem., 61:209, 1974.
49. Ashmead, D. and Graff, D., "Placental Transport of Chelated Iron," Proc. Int. Pig Vet. Soc. Congress, Mexico City, Mexico, p. 207, July, 1982.
50. Bagley, D. *et al.*, "Placental Transfer of Chelated Iron," Abs. No. 3, Fed. Proc., 25:189, 1966.
51. Beaconsfield, P., *et al.*, "The Placenta," Sci. Amer. 243:95, Aug., 1980.
52. Rajan, K., *et al.*, "Metal Chelates of L-Dopa for Improved Replenishment of Dopaminergic Pools," Brain Research, 107:317, 1976.
53. Agnew, L., *et al.*, *Dorland's Medical Dictionary* (Philadelphia: W.B. Saunders

Co.), 445, 1965.

54. Graff, D., "Radioactive Isotope Research with Chelated Minerals," in Ashmead, D., ed., *Chelated Mineral Nutrition in Plants, Animals and Man* (Springfield: Charles C Thomas), 275, 1982.
55. Christy, H., "The Use of Amino Acid Chelated Minerals in Egg Quality," in Ashmead, D., ed., *Chelated Mineral Nutrition in Plants, Animals and Man* (Springfield: Charles C Thomas), 196, 1982.

NAME INDEX

A

Adham, N., 82, 98, 101
Adibi, S., 148, 150, 151, 152, 170, 195
Agar, W., 119, 124, 170, 180
Agnew, L., 232
Albers, R., 111, 180
Albers, W., 111
Alcock, N., 101
Alfin-Stater, R., 179
Allison, B., 148
Alvardo, F., 204
Ames, B., 127, 147, 230
Anderson, C., 11
Andria, D., 131, 149
Antonov, V., 53
Antonson, D., 196
Ashmead, D., 180, 204, 212, 223, 231, 232
Ashmead, H. DeWayne, vii, viii, 28, 112, 125
Ashmead, Harvey H., vii, viii, 204, 229, 231
Ashmead, H., 147, 160, 181, 196

B

Bagley, D., 231
Bailar, J., 69
Bannerman, R., 124
Barbeau, A., 179
Barclay, G., 69
Barrett, A., 150
Barrowman, I., 52
Baxter, J., 69
Beaconsfield, P., 231
Beck, B., 231
Begwood, E., 195
Bell, C., 69, 70, 211 230
Bernard, Claude, 6, 12
Beutler, E., 102
Bezaeu, L., 99
Bezkorovainy, A., 170
Biber, B., 52
Bigwood, E., 147
Bjerrum, J., 101, 212
Black, E., 97
Blanc, B., 102
Blumberg, W., 102
Blunt, J., 28
Boyer, P., 12, 111
Bradley, H., 6, 12
Brambell, F., 148
Branton, D., 54
Brenner, F., 69
Brestcher, A., 53
Bretscher, M., 54
Brewer, G., 26, 52, 96, 196
Brickman, 84, 99
Briggs, F., 98
Briggs, G., 11, 22, 24, 28, 29, 52, 53, 96, 170
Bronner, F., 12, 95, 100, 195, 230
Brown, E., 96, 123, 124
Brown, M., 13, 26
Burston, D., 148, 151, 170, 196, 212, 231
Busch, D., 69
Bush, I.M., 179

C

Cabell, C., 27
Caggiano, 179
Callender, S., 28, 115, 124
Calloway, D., 11, 22, 24, 28, 52, 96, 170

Campbell, B., 100
Cantin, M., 99
Cardon, B., 159, 160, 169, 230
Cartwright, G., 52, 96
Case, R., 97
Cheeseman, C., 150, 195
Cheng, H., 53
Cheng, L., 28
Chenoweth, M., 69
Christy, H., 232
Church, A., 6, 12
Chutkow, 75, 96
Classen, H., 96
Code, C., 148, 195, 196
Code, C. F., 95, 230
Cohen, H., 99
Comar, C., 12, 69, 96, 100, 230
Conrad, M., 20, 28, 118, 124, 146, 152, 180
Corbes, F., 169
Cornell, J., 149, 151
Cott, Allan, 179
Cotzias, G., 3, 11
Cousins, R., 99
Craft, I., 148
Craig, D., 57, 69
Cramer, D., 95
Crampton, R., 102, 180
Crim, M., 100, 151, 195, 204, 212, 231
Crosby, W., 180
Curran, P., 150

D

Dalley, K., 147, 195
Dash, S., 179
Davidson, A., 52
Davies, B., 100
Davis, N., 27
DeGrazia, J., 101, 180
DeLuca, H., 28, 98
Dementzis, F. M., 179
Dick, A., 15, 27
Dickerson, J., 5, 12
Dinauer, R., 147, 170, 194, 230
Disler, P., 123
Donsbach, K., 97
Dowdy, R., 27
Drew, H., 56, 69, 160, 169
Dubach, R., 88, 89, 96, 100
Dwyer, F., 69, 70, 98, 147, 212, 229
Dyer, F., 211
Dyer, I., 16, 27

E

Earle, I., 27
Ebel, J., 96, 98, 194
Edwards, H., 101, 196
Egan, A., 27, 196
Eichhorn, G., 10, 12, 69
Elliott, K., 125, 148, 149, 150, 151, 152, 170, 181, 195, 196, 212, 231
Elliott, L., 147
El-Shobaki, F., 18, 28
Evans, G., 30, 51, 82, 83, 91, 95, 97, 98, 99, 101, 102, 180, 194, 204

F

Fang, Sen-Maw, vi, 125, 170, 180, 204
Farrahan, P., 229
Fein, J., 12
Fickel, T., 127, 147
Finean, J., 54
Flagg-Newton, J., 54
Fleischman, A.I., 28
Fleishman, M., 98
Florkin, M., 97, 100
Folsch, G., 128, 147, 195
Forstner, G., 53
Forth, W., 33, 51, 52, 75, 96, 97, 101
Fouad, T., 231
Freeman, J. A., 85, 99, 169, 195, 229
Friedman, M., 100
Fritz, J. C., 123
Fujita, M., 149
Funk, Casimir, 6
Furst, A., 69

G

Gallagher, C., 12
Garay, R., 112, 229
Garrahan, P., 112
Geer, J., 85, 99
Giese, A., 53, 54, 111, 181, 229
Gilvarg, C., 127, 147
Gitlin, D., 100
Gitler, C., 53
Glynn, I., 108, 109, 111
Goodhart, R., 12, 102, 151, 204, 212

Gower, I., 91, 101
Gozette, I., 150, 204
Grace, 83
Graff, Darrell J., vii, viii, 29, 54, 124, 147, 170, 181, 196, 203, 209, 210, 212, 223, 229, 231, 232
Granick, S., 100, 204, 230
Grassman, E., 100
Greer, J. C., 169
Greyton, A., 231
Guandalini, S., 130, 148, 170, 212
Guthrie, H., 5, 12, 24, 25, 28, 29, 52, 96, 101
Guyton, A., 52, 211

H

Hahn, C., 81, 98, 99, 101
Hallberg, L., 114, 116, 124
Hallbrook, 179
Hallman, P., 204
Hambidge, 179
Harper, H., 14, 26
Harrison, W., 181
Hastings, A., 98
Heaton, F., 84, 99
Hegsled, D., 195
Hemmings, W., 128, 148
Hemphill, D., 11
Henderson, P., 109, 112, 230
Henderson, R., 54
Hendrix, J., 99
Hendrix, T., 52, 89, 99, 100, 125, 148, 196, 229, 230
Henkin, R. I., 179
Henzel, J. H., 179
Hess, W., 170
Hill, C., 28
Hobbs, A., 111, 180
Hofez, E., 27
Hoffman, P., 111
Hoffman, W., 102
Holmes, R., 29
Horwitz, E., 69, 230
Hoskins, B., 69
Hsieh, S., 101
Hsu, Hsin-Hung, vi, 54
Hsu, J., 15, 26, 98, 101
Huebers, H., 97, 194
Hughes, M., 70
Hull, B., 53
Hunt, D., 100
Hutagalung, R., 181, 204

I

Ibata, Y., 178, 181

J

Jandl, J., 101
Jaya-Rao, K. S., 179
Jensen, N., 100, 169, 212, 230
Jodal, M., 38, 52
Johnson, B., 147, 195
Johnson, L., 53, 97, 111, 148, 150, 170
Jones, Evans, vi
Josefsson, L., 150, 151, 195
Justus, B., 96

K

Kagi, J., 99
Karcioglu, Z., 98, 101, 195
Karlish, S., 108, 109, 111
Kasai, M., 106, 111
Katz, J., 101
Kelly, D., 53
Kenny, A., 135, 150
Kerchgessner, M., 181
Kienholz, E., 99
Kim, Y., 134, 136, 148, 149, 150, 170, 195, 196
Kinter, W., 53, 149
Kirchgessner, M., 100
Kirkby, E., 54
Kleinzeller, A., 195
Knowles, A., 111
Kowalski, T., 99, 230
Kowarski, S., 111
Kramer, H., 52
Kramer, M., 52, 97, 150, 151, 194, 195, 204, 212, 231
Kramer, P., 111
Kratzer, F., 27, 170
Kuhn, I., 115, 123
Kumar, S., 179
Kummerow, F., 29

L

Larrson, I., 112
Lauterbach, F., 52, 97, 150, 151, 194,

195, 204, 212, 230
Layrisse, M., 117, 118, 119, 124, 161, 170
LeBlanc, F., 91, 101
Lebland, C., 53
Lee, G., 96
Levander, O, 28
Ley, H. Z., 63, 70
Liebig, 13, 15
Lodish, H., 54
Loewenstein, W., 54
Loubard, D., 149
Louvard, D., 149

M

MacInnis, A., 27, 124, 128, 148, 161, 170, 208, 209, 212, 217, 230
MacIntyre, I., 99, 101, 230
MacLennan, D., 111
Madara, J., 53, 54, 111, 150
Mahler, W., 169
Malagelada, J., 97
Maniloff, J., 100
Manis, J., 100, 124
Maroux, S., 149, 151
Martell, A., 69, 194, 195
Martinez-Torres, C., 118, 119, 124, 161, 170
Martonosi, A., 111
Masahiro, A., 111
Mason, K., 100
Matrone, G., 27, 28
Matthews, D., 131, 148, 149, 150, 151, 196, 202, 204, 212, 231
Meier, R., 91, 100
Meister, A., 139, 149, 151, 181, 195, 204, 212, 230, 231
Mellander, O., 127, 128, 136, 147, 148, 195, 230
Mellor, D., 69, 70, 98, 147, 211, 212, 229
Memmler, R., 51, 95
Mendel, L., 6, 12
Mengel, K., 54
Michell, R., 54
Miles, L., 231
Miller, A., 111
Miller, Gene W., vi, viii
Miller, R., 170
Miller, W., 52
Mills, C. F., 27, 100, 101, 113, 114, 123
Mitchell, E., 96
Miyamoto, H., 111
Moe, H., 52
Moeller, T., 69, 230
Montal, M., 53
Moog, F., 30, 51
Moore, C., 88, 89, 96, 100
Moore, Laurel, vi
Morgan, G., 56, 69, 160, 169
Mountcastle, V., 27, 125, 148, 196, 229
Moynahan, E. J., 179
Mukherjee, T., 53
Mulder, D., 16, 27
Munck, B., 150, 195
Munro, H., 100, 148, 151, 195, 204, 212, 231

N

Narins, D., 170
Nasset, K., 148, 204
Neutra, M., 53
Niccum, W. L., 123
Nicholas, D., 27, 196
Nicolaysen, R., 28
Nicolson, G., 53, 111, 194
Nightingale, R., 27
Noren, O., 151
Nyholm, E., 57, 69

O

Oberleas, D., 27, 196
O'Connor, M., 125, 147, 148, 149, 150, 151, 152, 160, 212, 231
O'Dell, B., 26, 100, 124
O'Connor, M., 181, 195, 196
Oelshlegel, F., 52, 96, 196
Ogata, E., 100, 204
Ostwald, T., 111
Otsuka, N., 178, 181
Owen, C., 96
Oxender, D., 112
Oyler, Dee, vi

P

Palekar, A., 151
Parsons, D., 150
Pauling, L., 229
Payne, J. W., 127, 147, 148, 149, 196
Pauling, L., 96, 97, 98

Persach, P., 102
Petering, H. E., 179
Peters, T., 133, 149, 196
Peterson, Robert V., vi
Petith, M., 99
Pfeiffer, C., 178, 179, 181
Pike, R., 13, 26
Plocke, D., 15, 29
Pollock, E., 54, 111, 195
Pories, W., 26, 179
Prasad, A., 26, 27, 51, 52, 96, 178, 179, 196
Praske, J., 13, 29

R

Racker, E., 111
Radhakrishman, A., 125, 137, 148
Raffin, S., 124
Rajan, K., 231
Rasmussen, H., 101
Reinhold, J., 28, 75, 96
Rhfeld, J., 112
Rhodes, J., 145, 149, 150, 212
Richards, M., 99
Riordan, J., 91, 101
Rish, M., 27
Robertis, E., 53, 54, 111, 181
Roberts, K., 52
Robertson, J., 53
Robinson, J., 204
Rose, R., 97
Rother, M., 123
Rothman, J., 54
Rothstein, A., 91, 93, 100
Rubin, W., 52
Rubino, A., 130, 148, 150, 170, 212
Rummel, W., 18, 28, 33, 51, 75, 96, 97, 194

S

Sahagian, B., 27, 93, 97, 101, 171, 180, 181, 194
Saroff, H., 97
Sarper, R., 98, 101, 195
Savage, J., 124
Scatchard, G., 97
Schachter, D., 51, 97, 100, 111, 124
Schade, S., 20, 28
Schedl, H., 99
Scheinberg, H., 74, 96
Scheiner, E., 101
Schepartz, B, 231
Schneider, H., 11
Schricker, B., 82, 83, 98, 194
Schroeder, H. A., 179
Schultz, S., 150
Schutte, K., 12, 26
Schwartz, A., 195
Schwartz, F., 181
Scott, M., 124, 170
Seelig, J., 54
Seelig, M., 99
Seigler, T., 124
Sell, 27
Semenza, G., 144, 151
Seven, M., 12, 69, 194
Sharpe, L., 96
Shils, M. E., 12, 102, 151, 179, 204, 212
Shoaf, C., 135, 150
Sigrist-Nelson, K., 139, 151
Silk, D., 131, 138, 149, 170
Sing, M., 101
Singer, S., 53, 54, 81, 111, 194
Sjostrom, H., 151
Skoryna, S., 27, 28, 51, 52, 53, 95, 96, 97, 99, 100, 101, 125, 180, 181, 194, 195, 196, 204, 229, 230
Smyth, D., 150, 195
Soffer, A., 69
Sollner, K., 44, 53
Solomons, N., 101, 195
Solvell, L., 114, 116, 124
Song, M., 82, 98
Specht, W., 52
Specian, R., 53
Spyker, D., 115, 116, 117, 118, 119, 124
Squibb, K., 99
Staehelin, L., 53
Stainhelin, L., 53
Starcher, B., 28, 100
Statz, E., 97
Steck, T., 54
Steenback, H., 28
Steiner, L., 69
Sternlieb, I., 74, 96
Stotz, E., 100
Subbuswamy, S., 52
Sullivan, 179

Suso, F., 101, 196
Suttle, N., 16, 20, 27, 196

T

Tansy, Martin, 85, 99, 230
Tate, S., 151, 195, 204, 212, 230, 231
Taylor, A., 73, 81, 95, 98
Taylor, T., 27
Templin, V., 28
Texter, E., 196
Thompsett, S., 51, 96
Thompson, A., 51, 97, 100, 101, 230
Thompson, D., 28, 196
Tiffin, L., 124, 147, 170, 194, 230
Tobias, J., 101
Tomsett, S., 97
Tosteson, D., 111
Trier, J., 53, 54, 111, 150
Triggle, D., 54
Turnball, A., 114, 124

U

Ugolev, A., 125, 132, 133, 138, 149, 150, 170, 204
Underwood, Eric J., 5, 12, 15, 24, 26, 28, 29, 97, 100, 101, 117, 124, 179, 204, 230
Unwin, N., 54

V

Valberg, L., 101
Vallee, V., 99
Van Campen, C., 96, 97, 99, 100, 123
Van Campen, D., 230
Van Deenen, L., 49, 54
Vannoti, R., 102
Vaughan, B., 52
Vohra, P., 27
Voorhees, 179
Votava, 83

W

Wacker, H., 144, 151
Waddell, D., 27
Waldron-Edward, D., 27, 28, 51, 52, 53, 95, 96, 97, 99, 100, 101, 125, 194, 195, 196, 204, 229, 230
Walker, N., 54
Wallace, A., 69
Warren, G., 54
Wasserman, R., 73, 79, 95, 98, 194, 230
Weber, K., 53
Weintraub, L., 124
Weir, E., 98
Werner, A., 56, 60, 70, 169
Wheby, M., 115, 116, 117, 118, 119, 124
Wheby, N., 115, 119, 124
White, A., 12, 54, 106, 111, 151, 161, 169, 180, 181, 195, 196, 204, 212, 229, 230
Whiting, F., 99
Widdowson, E., 5, 12
Williams, E., 128, 148
Williams, R., 12
Williams, S., 27, 125, 170
Wilson, D., 30, 54
Wilson, E., 12, 51, 52, 96, 97
Wilson, T., 53, 149
Windholz, M., 125, 148
Witt, W., 27, 28
Wood, D., 51, 95

Y

Yamamoto, M., 221, 231

Z

Zile, M., 98
Zull, J., 81, 98

SUBJECT INDEX

A

Absorption metal ions by active transport (*See* Active transport)
Absorption of peptides (*See* Peptides absorption in small intestine)
Absorptive cells
  binding of, 40, 41
    schematic diagram, 41, 42
  brush border of, 42, 43
    schematic diagram, 36
  desmosomes, 40-41
    schematic diagram, 36, 41, 42
  formation of, 35
  functions of, 35, 39
  fusing of, 40
  glycocalyx, 43
    schematic diagram, 41
  interior of, 42
    schematic diagram, 36
  location of, 35
  maturation process, 35
  microvillus (*See* Microvillus)
  mucosa histologic organization, schematic drawing, 38
  schematic diagram, 38
  tight junction of, 39-40, 41
    schematic diagram, 36, 41, 42
  true cells, 35, 38
Active transport, absorption metal ions by
  advantage over facilitated diffusion, 106
  calcium transport carrier, 106-107
  carrier proteins, 105, 214
    description, 105
    molecular weight, 105
  definition of, 105, 174
  energy supply for, 106, 214
  movement mechanism, 104-105
  movement of ion into cell, schematic diagram, 104
  mucosal cell membrane model, schematic diagram, 105
  of peptides, 138-139
  of non-ionized metals (*See* Secondary intestinal pathway for chelate absorption)
  sodium-potassium transport (*See* Sodium-potassium transport)
  source of energy, 107, 214
  summary, 110
Acetylcholine, involvement of in muscle contraction, 9-10
Alimentary canal (*See also* Small intestine)
  definition of, 30-31
  parts of, 31
  schematic diagram, 32
Aluminum, as nonessential contaminant, 4
Amino acid
  absorption through peptide pathway, 146
  amino acid chelate, schematic diagram, 79
  as chelating ligands
    characteristics of, 62
    common donor groups, 62
    glycine, 61
    metaproteins, 61-62
    of protein molecules, 61
  as found in nature, 63
    listing, table, 64
  biologic importance of, 63
  first studies of, 63
  formation five-membered chelate rings by, 63-64

Amine acid chelates dipeptide-like, absorption of (*See* Dipeptide-like chelates)
Arsenic, as nonessential contaminant, 4
Ascorbic acid, enhancement absorption non-heme by, 20
ATPase enzyme
as source of energy, 107
description, 108-109
mechanism of, 107
schematic diagram, 108
model, schematic diagram, 109, 110
molecular weight, 107

B

Barium, as nonessential contaminant, 4
Beryllium, as nonessential contaminant, 4
Bile, role of in intestinal uptake of metal ions, 77
Boron, as nonessential contaminant, 4

C

Cadmium, as nonessential contaminant, 4
Calcium
absorption of in duodenum, 30
effects of pH on, 73-74
rates of, graph, 74
site of, 78
sites of, diagram, 78
steps in, 87
summary, 94
as carrier protein, 105
as essential macromineral, 3, 11
chelation of to form heterocyclic ring, 184
schematic diagram, 184
intestinal pH and solubility of, 33
graph of, 34
involvement of in muscle contraction, 9-10
molecular weight calcium-binding protein, 183
need for structural strength, 6
percent absorbed in intestine, 24
percent in body, 4
recommended daily allowance, 22
role in intestinal absorption vitamin D, 20
theory of intestinal transport of, 79-81
need for vitamin D, 81
schematic diagram, 80
transport similarities other ions and, table 95
uptake of when chelated, 197
Calcium transport carrier
ATPase, 107 (*See also* ATPase enzyme)
binding capacity, 106-107
cause of transport, 107
description, 106
enzymatic reactions, schematic diagram, 110
physiology of, 107
role ATPase, 107
Carbohydrates
as body processes regulators, 8, 9
as source of energy, 7, 8
Carbon,
in body, 3, 11
percent in body, 4
Chelation
acidic and basic donor groups, 60-61
amino acid ligands (*See* Amino acid)
cisdihydroxytetrammine
cobalt (111) ion, 60
molecular structure, 60
coordinate covalent bond, 58, 215
covalent bond, 58
electrondeficient molecules, 59
multicenter bonding, 59
types of, 58-59
dipeptide-like copper glycinate chelate, molecular structure, 56
extreme ionic bond, 58
factors affecting stability of, 62-66, 162
characteristics metal ions, 66
coordination number, 66
coordination number, table, 67
nature of donor atoms, 65
number chelate rings formed with single metal ion, 63, 162
oxidation, 66
reduction and, 66
resonanic effects and, 64
role amino acids, 63-64
role central metal ion, 66
size chelate ring, 63
size of ion, 66
steric effect of ligand, 65
temperature, 66
factors governing bonding of an acid

group, 62
metal complex, molecular structure, 62
role pH, 62
first recognition of, 56, 160
five and six-member rings, 60, 61, 215
example, molecular structure, 61
stability of, 63, 215
formation of, 55, 214
example, 55
molecular structure, 56
need for heterocyclic ring, 55-56
formation dipeptide-like chelate, 55, 215
molecular structure, 56
formation double heterocyclic ring, 55
molecular structure, 56
formation ring-like molecule, 55
four member ring, 59-60
example, molecular structure, 60
stability of, 60, 63
industrial application, 160
ionic bonds, 57, 215
examples bonding, 57-60
induced attraction, 57
molecular structure, 57
static attraction, 57
ligands from one acidic and one basic group, 61
ligands used for, 161
metal
definition, 59
effect size of metal ion on, 59
illustration, 59
metal ligand bonds, 57
diagram, 58
molecular weight and absorption abilities, 161, 216
multicenter bonding, 59
postulations of Werner regarding, 56-57
requirement for, 55
requirements of ligand to chelate metal ion, 59, 213
requirements of polyvalent ion for, 79
role of, 216
to avoid zinc deficiencies, 179
solution stability of, 62
factors affecting, 62-63
source of word, 56
stability constants for, 161
stability of, 59-60
definition of, 62
factors affecting, 62-65
summary, 67
three-member rings, 59-60
use of term, 160, 205-206
variation size of ring, 59-60
example, 60
with amino acids, 120, 126
Chlorine
as essential macromineral, 3, 11
percent in body, 4
Chromium
as essential micromineral, 4
percent in body, 4
Cobalt
cisdihydroxytetrammine cobalt (111)ion, 60)
molecular structure, 60
hexammine-cobalt (111)formation, 57
percent in body, 4
Copper
absorption of
ionized, 30
non-ionized, 113-114
as essential micromineral, 4
comparison uptake chelate to, 163
table, 163, 164
dipeptide-like copper glycinate chelate
formation, schematic diagram, 216
intestinal uptake, table, 121
molecular structure, 56
effects intestinal pH on absorption of, 74
diagram, 78
formation dipeptide-like chelate molecular structure, 56
history need for, 6
intestinal pH and solubility of, 33
graph of, 34
intestinal uptake of, 121
results study of, table, 121
table, 167
percent absorbed in intestine, 25
percent in body, 4
transport and absorption of,
chelation of, 91
copper-binding proteins, 91
ligand behavior and pH, 91
properties of copper-binding proteins, table, 92

similarity to zinc, 91
steps in, 87-90
summary, 92, 94
transport similarities other ions and, table, 95
uptake of when chelated, 197

D

Desmosomes
description, 41
location, 40
schematic diagram, 36, 41, 42
Dipeptide-like chelate intestinal absorption as small peptides (*See* Small peptides)
Dipeptide-like amino acid chelate
absorption of
analysis findings, 120-123, 127
evidence for transport of, table, 132
into microvilli, 154
into mucosal cells, 131
microvilli, magnification, 157
of methionine, 129
procedures used for, 119-120
results studies of, table, 121
study pathways for, 119-123
comparison uptakes chelates to mineral salts, 163
table, 163, 164
concept of, 126
theoretical, schematic diagram, 127
x-ray diffraction, 127
formation, 215-216
of copper, schematic diagram, 216
intracellular hydrolysis of (*See* Intracellular hydrolysis)
molecular weight calcium-binding protein, 182
molecular weight of, 121
mucosal cell transport of as peptides (*See* Mucosal cell transport)
phosphorylated peptides, 128
proposed membrane transport of, 217-218
schematic diagram, 218
requirement for uptake of, 142-143
role in hydrolysis of peptides, 144
site hydrolysis of, 130-131
"smuggler" transport concept, 127-128, 136
studies of, 127-128
transport mechanisms, 130-131, 217-218
evidence for, 130
uptake system of, 142
Dipeptide-like chelate locus determination (*See* Locus determination)
Duodenum
absorption of calcium in, 74, 78
diagram, 78
absorption of copper in, 74, 78-79
diagram, 78
absorption of iron in, 30
absorption of magnesium in, 74-75, 78
diagram, 78
absorption of manganese in, 76
absorption of zinc in, 30, 75-76
diagram, 78
absorption of zinc in, 30, 75-76
definition of, 73
length, 31, 73
pH of, 31, 73
role pancreatic fluid in, 77
solubility iron in, 75, 78
diagram

E

Enzymes
catalytic activity of, 11
chelation of, 10
definition of, 10
role of in mineral function, 10

F

Fats
as source of energy, 7, 8
effects on intestinal absorption minerals, 21-22
Fluorine
as essential micromineral, 4
percent in body, 4
Food, physiological role of, 5

G

Glutathione
attachment chelate to, 217
schematic diagram, 218
proposed member transport of, schematic diagram, 218

Hemoglobin
alternative pathways for, 117
effect histidine on absorption, 118
heme as primary source absorption, 116
molecular weight heme molecule, 117
molecular weight of, 146
role spleen, 220
studies uptake of, 114-115, 117
study of in mineral deficient animals, 220
hemoglobin levels, graph, 221
Histidine
mucosal uptake of, 199-200
results study, table, 201
mucosal-to-serosal transport of, 201-202
table, 201
Hydrogen
in body, 3, 11
percent in body, 4

I

Ileum, 31
Interaction Chart of Mulder, 16
illustration, 16
Intestinal absorption dipeptide-like chelates as small peptides (*See* Small peptide intestinal absorption
Intestinal absorption metal ions and chelates
absorption amino acids by mucosal cells, 217
action on mineral ingested
basement membrane action, 214
chelation mineral, 213-214
effect pH on, 213
in duodenum, 213
in stomach, 213
mucosal cell action, 214
transport mechanism, 214-215
conclusions, 228-229
enzymatic steps transport chelate, 218
final release of chelate into cytoplasm, 218
formation chelates, 216
formation dipeptide-like amino chelate, 215
formation heterocyclic rings, 215
of copper, schematic diagram, 216
glutathione attachment to chelate, 217
heterocyclic ring formation, 215-216
ionic bonds, 215
membrane transport of chelate, schematic diagram, 218
molecular weight and absorption molecule, 216-217
peptide hydrolysis, 218-219
release metal as free ion, 219
role spleen, 220-221
secondary process of, 219-220
implications of, 220
secondary transport mechanism requirement to function, 215
"smuggle" concept, 215
source covalent bond, 215
study fetal uptake of iron, 226
results, table, 227
study incorporation iron in RBC, 221
iron levels, graph, 222
study L-Dopa in brain, 227
study mineral deficient animals, 220
hemoglobin levels, graphs, 221
study serum ferritin concentrations, 221-222, 223, 224, 225, 227
hematocrit values, table, 223
hemoglobin concentration changes, graph, 225
hemoglobin concentrations, table, 223
procedure used, 223-224
regressive equations hemoglobin concentrations, graph, 225
results of, table, 222
serum ferritin concentration changes, graph, 226
serum ferritin concentrations, table, 224
study using dipeptide-like chelates, 227-228
implications findings of, 228
metabolism calcium, table, 228
transport system and form mineral ingested, 213
active transport (*See* Active transport)
diffusion transport, 213
Intestinal absorption minerals
absorption rates and region small intestine, 23
diagram, 24
and capacity cell membrane to bind, 23-24
illustrations, 21-25

control calcium absorption by hormone, 20
diagram, 21
dietary phosphates and, 18
effects ascorbic acid on absorption iron, 20-21
diagram, 22
effects fats on, 21-22
effects fiber on, 23
effects pH on solubility ferric iron, 20-21
diagram, 22
factors affecting metal absorption, 17-18
in diet, 17-18
in digestive tract, 17, 18
implications research findings, 213-229
importance pH of intestinal lumen, 23, 213
in vitro studies using rats, 25
results, table, 14
interference with by other minerals, 20
mineral antagonism
from salts in intestine, diagram, 17
mechanism, 16
mineral interactions influencing
categories, 16-17, 18
competition for same transport carriers, 18
concept used, 16
effects chemical form of mineral, 20
formation of insoluble complexes, 17
metabolic pathways, 19
metal component change of metallo-enzyme, 19
multiple interactions and effects of, 20
synthesizing of metal-binding proteins, 19
mucosal cell membrane, 213-214
role ascorbic acid on absorption non-heme, 20
solubility mineral salts, 23
summary studies of, 25-26, 213-229
transport systems, 213
use Interaction Chart, 16
antagonism from salts, diagram, 17
Intestinal uptake of free metal ions (*See also* Small intestine)
absorption of
calcium (*See* Calcium)
copper (*See* Copper)
iron (*See* Iron)
magnesium (*See* Magnesium)
manganese (*See* Manganese)
zinc (*See* Zinc)
absorption sites, diagram, 78
amino acid chelate, schematic diagram, 79
barriers for metal from lumen to blood, 71-72
basement membrane, 73
basolateral plasma membrane, 72-73
microvilli, 71-72 (*See also* Villus)
basolateral plasma membrane
description, 72-73
movement nutrients through, 73
tight junction of, 72-73
calcium-binding proteins, 79-81
functions of, 81
chelation requirements, 79
effects pH on solubility and absorption rates
on calcium, illustration, 73-74
on copper, 74
on iron, 75
on magnesium, 74-75
on manganese, 76
on zinc, 75-76
maintenance pH and, 77-78
microvillus (*See* Microvillus)
mucosal cell membrane transport mechanisms, 94
similarities, table, 95
need for binding capability of, 79
need for mineral ionization, 73, 213
need for mineral solubilization before absorption, 73, 213
of manganese (*See* Manganese)
problems in research on, 71
role of bile in, 77
role pancreatic fluid in, 77
route of nutrients to blood, 72-73
summary, 94
theory intestinal transport calcium, 79-81
need for vitamin D, 81
schematic diagram, 80
tight junction
location, 72-73
schematic diagram, 72
zinc-binding ligands, 81-82 (*See also* Zinc)

Intestine (*See* Small intestine)
Intracellular hydrolysis of dipeptide-like chelate, 205-211
chelates of iron made, 207
dissociation chelate in tissue, 210
intracellular hydrolysis and, 210
effect time in cell on, 211, 218-219
site of, 211
mucosal uptake and transfer chelated iron, 207-208
table, 208-209
procedures used in study, 206
purpose of study, 205, 206
reports research on, 208-209
role cellular pH in, 210
role peptidase enzymes in, 205
summary of, 211
Iodine
as essential micromineral, 3
percent in body, 4
Iron
absorption ferritin into mucosal cells, magnification, 158
absorption into microvilli, magnification, 157
absorption into mucosal cells, 154-155
absorption of
effects ascorbic acid on, 20-21, 114
in ferrous form, 75-76, 85-89, 114, 214
site of, 30
as essential micromineral, 4
comparison uptake chelates to, 163
table, 163, 164
concept absorption, 153-154
schematic representation, 154
effects pH on solubility, 20-21, 33, 114
diagram, 22
graph of, 34
ferritin aggregates in apical cytoplasm, magnification, 156
forms of iron for mucosal uptake, 115-116
hemoglobin iron absorption (*See* Hemoglobin)
history need for, 6
intestinal uptake of, 120-122
graph, 169
results study, table, 121
iron carbonate from lumen, magnification, 155
mucosal uptake and transfer chelated iron, 207-208
table, 208, 209
percent absorbed in intestine, 24-25
percent in body, 4
recommended daily allowance, 22
relationship absorption to chelation, 146-147
solubility and absorption of
diagram, 78
in duodenum, 75, 78
in duodenum, graph, 76
in ileum, graph, 76
in jejunum, graph, 76
method study of, 75
sources iron absorption, 117
storage and release of in mucosal cell, diagram, 88
transport and absorption of
apoferritin and, 87-88
binding on cell membrane, illustration, 86
movement, diagram, 90
movement of ferritin, 88
neutralization of ions, 85
osmotic pressure as transport, 88-89
steps in, 87
summary, 89-90, 94
transport of iron, 86-87, 214
transport similarities other ions and, table, 95
uptake when chelated, 197

J

Jejunum
absorption iron in, 30
pH of, 31
use of term, 31

L

Law of the Minimum
definition of, 13, 15
illustration, 14-15
Lead, as nonessential contaminant, 4
Liebig's Law, 13, 15
Lipid
as body processes regulators, 8, 9
for growth and tissue maintenance, 8

Lithium, as nonessential contaminant, 4
Locus determination for intestinal absorption of dipeptide-like chelate
absorption sites, 190-191
cellular dissociation of chelate, 186
chelate differences, 185
chelation calcium to form heterocyclic ring, schematic diagram, 184
conclusions from study, 191-194, 220-221
differences between chelates, 185
dipeptide-like amino acid chelate structure, x-ray diffraction, 184
enzymatic reactions, 185
gamma ray spectrophotometry results, table, 188
importance molecular weight, 184
integral protein, schematic diagram, 184
ionic pathway to absorption
chelation calcium, schematic diagram, 184
chelation of cation, 183-184
composition binding protein, 184
dipeptide-like amino acid chelate, x-ray diffraction, 184
integral protein in mucosal cell membrane, schematic diagram, 183
solubilizing mineral, 182
procedures used to study, 186-188, 220
use serial blood samples, 187-188
use zinc isotopes, 187
results study absorption zinc, 188-190
graph, 189
table, 188
two step absorption pathway, 185-186

M

Macromineral
essential, defined, 3
those essential for life, 3, 11
Magnesium
absorption of
in small intestine, 30, 121
ionization of, 85
similarity to calcium, 84
site of, diagram, 78
steps in, 87, 94
transport of, 85, 94
as essential macromineral, 3, 11
comparison uptake chelates to, 165
table, 163, 164
intestinal pH and solubility of, 33, 74-75, 78
diagram, 78
graph of, 34
intestinal uptake of, 121, 166
graph, 166
results study, table, 121
percent absorbed in intestine, 24
percent in body, 4
transport similarities other ions and, table, 95
uptake of when chelated, 197
Manganese
absorption in duodenum, 30, 76, 78
diagram, 78
steps in, 87
as essential micromineral, 4
intestinal pH and solubility of, 33
graph, 34
percent absorbed in intestine, 25
percent in body, 4
transport and absorption of
effect other metal ions on, 93-94
site of bonding, 93
steps in, 92-93
transport similarities other ions and, table, 95
Mercury
as nonessential contaminant, 4
inhibitor of cytoplasmic enzymes, 145
Metabolic energy
source, 106
use, 106
Metal ion antagonism, cause of, 7
Metalloenzyme
definition of, 10
metal component of change, 19
Microminerals
essential for life, 3-4, 11
those unproven as essential for life, 4
value of, 4-5
Microvillus
calcium-binding protein in, 81
contents core of, 72
cytoplasm defined, 72
description, 71-72
function of, 35, 43
glycocalyx, composition of, 43

integral proteins, 45-48
arrangement of, 47
cross-section, 50
model, schematic diagram, 46
movement across lipid bilayer, 46
proposed model of, 49
ratio lipid to, 47
representative structure of, 48
lipid molecules in, 43-44
and protein ratios, table, 49
bilayers of, 44
fluid mosaic model, 45
protein molecules in, 44-46
schematic diagram, 45
structure of, 43-45
membrane proteins
integral proteins, 45-47
lipid and protein ratios in membranes, table 49
peripheral proteins, 45
mucosal cell with, schematic diagram, 72
peripheral proteins, 45
permeability of, 49
schematic diagram, 50
plasma membrane of
characteristics of, 43
cross section of, 46
fluid mosaic model, 45
glycocalyx of, 43
intracellular dense layer, 43-44
layers of, 43-44
lipid molecules in, 43-44
luminal layer, 43
permeability properties of, 44-45
proposed model of, 49
proteins of, 44, 45-46
representative structure, 48
schematic diagram, 45
protein molecules in lipid bilayers, 44
protein to lipid ratio in, 47
table, 49
schematic diagrams, 36, 41
Milieu interieur, defined, 5
Mineral absorption from intestine, 13-16 (*See also* Intestinal absorption of minerals)
Mineral elements, body composition of, schematic diagram, 4
Mineral needs of body, 3-11
body elements composition
effect changes in, 5, 11
schematic diagram, 4
percent body atoms of, 3
percent body weight of, 3
Minerals
as body processes regulators, 8, 9
as source of energy, 7, 8
enzymatic co-factor function, 8-9
essential defined, 3
for growth and tissue maintenance, 8
function of, 6
history need of, 5-6
interactions of, 13-14, 15
first studies of, 15-16
illustration, 14-15
in plants, 16
interaction chart, 16
macromineral (*See* Macromineral)
micromineral (*See* Micromineral)
problems interpreting research results, 7
causes, 7
research on, 6-7
problems interpreting results, 7
types of, 6-7
trace minerals required by body, 4, 11
types research on functions of, 6-7
Molybdenum
as essential micromineral, 4
percent in body, 4
Mucosal cell membrane
description, 103, 213-214
fluidity of, 104
model, schematic diagram, 105
transport mechanisms, 94
similarities, table, 95
Mucosal cell transport of dipeptide-like chelates as peptides
enzymatic reactions required for ion transfer, 198
improved absorption due to chelation, 197
intestinal absorption peptides and amino acids, 199-200
schematic diagram, 199
intracellular hydrolysis in, 202
location of, 202
mucosal uptake histidine, 200
table, 201

mucosal-to-serosal transport, 201-202
table, 201
pump needed for membrane transport ion, 198
purpose study of, 198, 220
role spleen, 220
study techniques used, 199, 220
summary study findings, 203, 205, 220-221
uptake calcium, 197
uptake copper, 197
uptake iron, 197
uptake magnesium, 197
uptake zinc, 197
zinc absorption in mucosal cell, 200
results study, table, 201
Mucosal uptake metal ions
study of
analysis findings, 120-123
chelates used, 120, 121, 122
metal salts used, 120
procedures followed, 119-120
results analysis, table, 121

N

Niacin, role in protein synthesis, 14
Nickel
body need for unproven, 4
percent in body, 4
Nutrient metabolism
Law of the minimum and, 13-14
summary of, diagram, 14
Nitrogen
in body, 3, 11
percent in body, 4
Nutrients
classification of, 7
table of, 8
definition of, 5
for growth and maintenance of tissue, 8
table, 8
function of, table, 8
groups of essential, 5
sources of energy, 7-8
table, 8

O

Oxygen
in body, 3, 11
percent in body, 4

P

Pancreatic fluid, role in solubilization nutrients, 77
Peptide absorption in small intestine, 126-147
absorption processes, 132-133, 217
absorption sites, 129, 217
absorption without hydrolysis, 145
active transport mechanism of, 138-139
schematic diagram, 139
brush border enzyme, 134
calcium-peptide chelate and, 136
chelation with amino acids, 126, 217
cytosal enzymes, 134
digestion dietary protein, table, 129
dipeptide-like amino acid chelate (*See* Dipeptide-like amino acid chelate)
enzymatic systems, 134
enzyme reactions, steps in, 140-142
enzymes in enterocyte, table, 144
evidence intracellular hydrolysis of, table, 132
evidence transport of, table, 132
glutathione movement, 139-140
hydrolysis oligopeptides into amino acids, 130
hydrolysis protein, 129-130
effect pH on, 129-130
intracellular hydrolysis, 137
intralumen hydrolysis, 137
intramembrane hydrolysis, 137
membrane hydrolysis, 130-132
evidence for, table, 133
molecular size and transport of, 131
molecular weight cytosol enzymes, 145
movement of across cell membrane, 139-140
schematic diagram, 140
mucosal cell transport of dipeptide-like chelates as (*See* Mucosal cell transport)
peptidase activity in villus, 134
peptidases in brush border membrane, 135-136
effects metals ions on, 136
peptidases in kidney brush border proximal tubules, 135
peptide hydrolysis, diagram, 137

principal site for digestion luminal peptides, 130, 217
protein sources, 129
purification amino-oligopeptidases, 134-135
relationship food on, 217
structural requirements for transport of, 143
features of, 143
schematic diagram, 143
subcellular distribution peptidases, 134
table, 135
summary study of, 146-147, 217-221
superficial hydrolysis, 137
transport free amino acids, 137-138
transport mechanism of, 130-131, 220
free amino acids, 138
peptides, 138
small peptides, 138
Phosphorus
as essential macromineral, 3, 11
need for structural strength, 6
percent in body, 4
Phosphorylated peptides, 128
Potassium
as carrier protein, 105
as essential macromineral, 3, 11
percent in body, 4
sodium-potassium transport (*See* Sodium-potassium transport)
Prosthetic groups of enzymes (*See* Metalloenzyme)
Protein
as body processes regulators, 8, 9
as source of energy, 7, 8
for growth and tissue maintenance, 8
Protein synthesis, nutrients involved in, 14-15

R

Rubidium, as nonessential contaminant, 4

S

Science of nutrition, definition of, 5
Secondary intestinal pathway for chelate absorption
advantages over ionic mineral absorption, 219-220
implications of, 220
chelates used in study of, 120, 121, 122
metal salts used in study, 120
procedures followed to study, 119-120
requirements for pathway, 153
results,analysis of, 120-123, 153
table,121
transport found, 122
Selenium
as essential micromineral, 4
percent in body, 4
Silicon
percent in body, 4
unproven body need for, 4
Small intestine (*See also* Intestinal uptake free metal ion)
absorption rate minerals in, 32-33
absorptive function, summary, 50-51
alimentary canal and, 30-31
assimilation of nutrients begun, 30
circular folds of, 33-35
schematic diagram, 36
coiling of, 33
schematic diagram, 34, 37
duodenum, 31 (*See also* Duodenum)
effect pH on solubility of minerals, 33
graph, 34
entry to, 31
exit from, 3
goblet cell, 38
histological organization mucosa, 38
layers of, 34-35
length of, 31
minerals absorbed in, 30
mucosa of digestion and absorption
absorptive cells (*See* Absorptive cells)
schematic diagram, 36, 37, 38, 39
villus (*See* Villus)
pH of, 31
requirements for absorption nutrients, 33
secondary intestinal pathway for chelate absorption (*See* Secondary intestinal pathway for chelate absorption)
sections of, 31
solubility and absorption minerals in, 31-33
structures of
circular folds,, 33, 34, 36
coiling, 33, 34
layers, 34-35

schematic diagram, 36, 37
villus (*See* Villus)
Small peptide intestinal absorption
active transport defined, 174
binding sites, 172
enzymatic reactions carrier protein, 176
schematic diagram, 177
intestinal transport zinc, 175
graph, 175
need ATP for absorption, 175
need of body for zinc, 178 (*See also* Zinc)
practical application study findings, 177-178
source metabolic energy, 174
studies of, 171
study procedures, 172-174
handling segments, diagram, 173
summary, 179-180
transport metal ion, 176
two-step process for, 171
uptake of chelate, hypothesis, 172
"Smuggler" concept transport mechanism, 127-128
Sodium
as carrier protein, 105
as essential macromineral, 3, 11
percent in body, 4
sodium-potassium transport (*See* Sodium-potassium transport)
Sodium-potassium transport, 107
dissociation constant for, 109
enzymatic reactions model, schematic diagram, 110
mechanism of, 108-110
pump affecting ATP hydrolysis, schematic diagram, 109
pump in cell membrane, schematic diagram, 108
Stomach, pH of, 73, 77
Strontium, as nonessential contaminant, 4
Sulfur
as essential macromineral, 3, 11
percent in body, 4

T

Tin
percent in body, 4
unproven body need for, 4
Trace minerals
essential for life, 4, 5
importance metabolic interactions of, 15
Transport systems of digested mineral, 213
Two mineral absorption pathways
absorption ferritin into mucosal cells, 154
magnification, 158
absorption iron dipeptide-like amino acid chelate, 154, 155-156
magnification, 157
absorption iron into mucosal cells, 154-155
magnifications, 155, 156
chelate absorption model requirements, 156-157
conclusion of study, 163, 164, 167-168
premises for, 164-166
controversies regarding, 159-160
dipeptide-like amino acid chelate absorption, 157-158, 162-167 (*See also* Dipeptide-like amino acid chelate absorption)
uptake compared to mineral salts, table, 162, 164
effect pH on metal uptake, 164
fate chelates after mucosal cell absorption, 158-159, 163
ferritin aggregates in apical cytoplasm, magnification, 156
intestinal mineral absorption *in vitro*, 157-158, 162 (*See also* Dipeptide-like amino acid chelate)
intestinal uptake of copper, graph, 167
intestinal uptake of iron, graph, 169
intestinal uptake of magnesium, graph, 166
intestinal uptake of zinc, table, 168
ionic mineral absorption concept, 153-154
schematic representation, 154
iron carbonate from lumen, magnification, 155
possible fates ingested dipeptide-like amino acid chelate, schematic diagram, 159
radioactive isotope *in vivo* studies, 158, 168 (*See also* Locus determination intestinal absorption dipeptide-like chelates)
role form of mineral, 213
types studies of, 157-158

use chelation to study, 160
use EDAX readings to study, 154-155

V

Vanadium
percent in body, 4
unproven body need for, 4
Villus
absorptive cells on, 39
anatomy of, 38-40
description, 35
goblet cell of, 38
schematic diagram, 37
location, 35
microvilli, 35
peptidase activity in, 134
schematic diagram, 36, 37, 38, 39
vascular anatomy of, 38-39
schematic diagram, 39
Vitamins
as body processes regulators, 8, 9
as source of energy, 7, 8
discovery of, 6
for growth and tissue maintenance, 8
role in intestinal absorption calcium, 20
use of term, 6
Vitamin D, need for to form calcium-binding protein, 81

Z

Zinc
absorption into mucosa cell, 185
absorption through duodenum, 30, 73
steps in, 87
activator for S H enzymes, 145
as complex peptide, 182
molecular weight, 182
as essential micromineral, 4
chelation of in mucosal membrane, 82
schematic representation, 82
clinical disorders from deficiency of, table, 179
comparison uptake chelates to, 163
table, 163, 164
cysteine amino acid chelates of, molecular structure, 61
effects deficiency of, 14-15
history need for, 6
hypogonadism due deficiency of, 178
influence different proteins on absorption of, 117-118
inhibitor of cytoplasmic enzyme
intestinal pH and solubility of, 33, 75-76, 78
graph, 34
schematic diagram, 78
intestinal transport of, 175
graph, 175
intestinal uptake of, 120-122
graph, 168
results, table, 121
molecular weight zinc-binding protein, 183
mucosal uptake of, table, 201
mucosal-to-serosal transport, 201-202
table, 201
need for by brain, 178
niacin function and, 14, 15
percent absorbed in intestine, 24
percent in body, 4
ratio ionic zinc to chelate zinc, 202-203
role in DNA, RNA and growth, 178
role in protein synthesis, 14-15
study zinc chloride absorption, 186-188
results of, graph, 189
results of, table, 188
study zinc dipeptide-like chelate absorption, 186-188
conclusions, 191-194
results, graph, 189
results, table, 188
transport similarities other ions and, table, 95
uptake of when chelated, 197
zinc-binding ligand, 81-82, 94
albumin protein of, 83-84
amino acids in, 82
bipolar nature of, 82-83
location of, 83
results study of in rats, table, 84